Elementary General Relativity

C. Clarke

Lecturer,
Department of Mathematics,
University of York

A HALSTED PRESS BOOK

JOHN WILEY & SONS
New York

First published 1979
by Edward Arnold (Publishers) Ltd.
Published in the U.S.A. in 1980
by Halsted Press, a Division of
John Wiley & Sons, Inc., New York

British Library Cataloguing in Publication Data

Clarke, C.
Elementary general relativity.
1. General relativity (Physics)
I. Title
530.1'1 QC173.6

ISBN 0-470-26930-8

Cover picture courtesy of
the Royal Greenwich Observatory

Printed and bound in Great Britain
at The Pitman Press, Bath

Preface

This book provides a brief and systematic account of Einstein's general theory of relativity, incorporating the coordinate-free, geometrical methods that have given relativists in recent years a much deeper insight into the meaning of the theory. General relativity brings together profound physical ideas, an elegant geometrical structure, and the computations with differential equations familiar in traditional applied mathematics. My aim has been to give adequate weight to each of these while staying within the scope of a genuinely elementary book, suitable for a student who does not necessarily want to specialize in the subject or to devote himself to an exhaustive study of all its aspects. I assume on his part a knowledge of a little basic calculus (simple differential equations, partial derivatives and, ideally, a glancing acquaintance with the wave equation) and some linear algebra (vectors and matrices, and the idea of a vector space).

The plan of the book is as follows.

Chapter 1 develops special relativity in a setting and notation that can immediately be transferred to general relativity. Most of the fundamental geometrical ideas are established here.

Chapter 2 gives a more conventional account of some selected applications of special relativity. Only a few results, contained in the first two sections of the chapter, will be quoted in later chapters, so that the reader already familiar with the special theory may omit the whole of this chapter.

Chapter 3 is the heart of the book. A geometrical model of space–time is progressively built up, motivated by physical arguments stemming from the equivalence principle, leading to Einstein's field equations.

Chapter 4 deals very quickly with the simplest form of weak-field theory, with application to gravitational radiation.

Chapter 5 concludes the book with a fairly detailed analysis of the Schwarzschild solution, plane fronted gravitational waves, and the Robertson–Walker cosmological solutions.

Exercises at the end of each chapter extend the general theory into particular applications, giving a broader picture of the scope of the subject.

Notation

An index to particular symbols and letters will be found at the end. Apart from these, the following general conventions are used.

An equation with a colon, such as

$$P := \ldots$$

is a definition. The left-hand-side (here 'P') is defined to be whatever is on the right-hand-side.

Ordinary boldface letters (**x, y, a, b, X, Y, L,** etc.) denote four-dimensional column vectors, row vectors, 4×4 matrices etc. Sans serif letters (**x**, **y**, **r**, **R**, etc.) denote three-dimensional row matrices etc.

Four-dimensional *geometrical* vectors (as opposed to the column matrices formed out of their components in some coordinate system) are denoted by ordinary italic or Greek letters, usually X, Y, Z etc.

The four coordinates of space–time are numbered from 0 to 3, the zeroth one being timelike. The signature of the metric is $(-, +, +, +)$.

Equations are numbered separately in each chapter. Thus (16) refers to equation (16) in the current chapter. Chapters are divided into sections and subsections. Thus '2.3' denotes section 3 of chapter 2, while '2.3.4' refers to its fourth subsection.

York
1979

C.C.

Contents

Index of symbols

1
Synopsis of special relativity

1.1 Coordinates and inertial observers

It is a basic assumption of Newtonian mechanics that whenever one is dealing with physical space one can introduce into it Cartesian coordinates so as to express the laws of motion in a standard simple form. For instance, the law governing the motion of a particle of mass m that is attracted gravitationally to another particle of mass M is given by assigning coordinates $\mathbf{r} := (x^1, x^2, x^3)$ to the first particle and similar coordinates $\mathbf{r}_0$ to the second, so that the law becomes

$$m\frac{\mathrm{d}^2}{\mathrm{d}t^2}\mathbf{r} = \frac{GMm(\mathbf{r}_0 - \mathbf{r})}{|\mathbf{r}_0 - \mathbf{r}|} \tag{1}$$

Usually no emphasis is placed on the exact nature of these coordinates, but if pressed one could describe how they might be set up by some kind of measuring apparatus. Note that in this Newtonian picture one has a three-dimensional Euclidean space S (the set of all 'places') and a quite separate time T (the set of all 'instants') which are given coordinates (x^1, x^2, x^3) and t respectively; at each instant of T a particle is at a certain place in S.

Relativity is formulated in terms of a four-dimensional *space–time* M, which is thought of as the set of all places at all times. In special relativity it can be given coordinates (t, x^1, x^2, x^3). A particle occupies a path in space–time called its *world-line:* the set of all its positions at all times. When one wants to distinguish the points of M from spatial points in S one calls those in M *events.*

In special relativity and in the modern form of Newtonian theory (though not for Newton himself!) there is no one coordinate system that forms an absolute reference frame which can be selected at the outset. Instead there is a special class of coordinate systems, with respect to each of which the equations of motion take on a standard simple form (such as (1)). Coordinate systems in this class are called *inertial coordinates* or *inertial frames.* Corresponding to this, we assume that there is a special class of observers, inertial observers, who on following some standard procedure can establish inertial coordinates and find the standard forms of the laws of motion satisfied. One can think of these as uniformly moving observers: more precisely, if O is an inertial observer and if O′ is some other observer who is moving uniformly according to the coordinates of O, then O′ will also be an inertial observer; if O′ is accelerating or rotating, according to O's inertial frame, then he will not be inertial.

The concept of the inertial observer is valid in relativity theory as well, and for both special relativity and Newtonian theory we have the following:

Principle of Relativity: The laws of physics are the same for all inertial observers.

This principle was first clearly annunciated by Galileo, when he wanted to show that the rapid motion of the Earth, according to the controversial Copernican theory, is not noticed by us because it is almost uniform. He illustrated the principle by considering experiments performed by two imaginary observers, one at rest on a quay and the other shut up in the cabin of a ship moving steadily over calm seas. Each would observe precisely the same phenomena: for instance, if either let a body drop from an initial state of rest relative to himself then he would see it drop vertically relative to himself, even though it would seem to be moving in a parabola relative to the other observer.

1.2 Coordinate transformations

A coordinate system in special relativity is, by definition, a one–one correspondence between points of space–time M and the set $\mathbb{R}^4$ of ordered quadruplets of numbers (x^0, x^1, x^2, x^3) (where we now introduce

$$x^0 := t$$

to make our notation more coherent).

If we have two sets of coordinates, (x^0, x^1, x^2, x^3) and $(x^{0\prime}, x^{1\prime}, x^{2\prime}, x^{3\prime})$, then it will be possible to express the x' as functions of the x and vice versa. We now introduce a matrix notation to discuss this functional relationship.

$\mathbf{x}$ will denote the column matrix $\begin{pmatrix} x^0 \\ x^1 \\ x^2 \\ x^3 \end{pmatrix}$

The column matrix $\mathbf{x}$ depends on the point x of space–time whose coordinates are being considered, so we write $\mathbf{x} = \mathbf{x}(x)$. (Note that $\mathbf{x}$ denotes a column, x denotes a point and x^0 etc. denote components.)

The vital property possessed by inertial coordinates is that for any two such coordinate systems there is a non-singular 4×4 matrix $\mathbf{L}$ and a constant column vector $\mathbf{a}$ such that

$$\mathbf{x}'(x) = \mathbf{L}\mathbf{x}(x) + \mathbf{a} \qquad \text{for all } x \in M \tag{2}$$

and conversely

$$\mathbf{x}(x) = \mathbf{L}^{-1}\mathbf{x}'(x) - \mathbf{L}^{-1}\mathbf{a} \qquad \text{for all } x \in M \tag{2'}$$

We can now use this to give the principle of relativity more mathematical content. The principle implies that if we take a system of equations expressing a law of physics in terms of coordinates $\mathbf{x}$, and we substitute for $\mathbf{x}$ its expression (2′) to give the law in terms of $\mathbf{x}'$, then the resulting equations are the same as those obtained by simply replacing $\mathbf{x}$ by $\mathbf{x}'$. This property is called the *invariance* of the equations under (2). (This might be thought false for vector equations such as

$$m\mathrm{d}^2\mathbf{r}/\mathrm{d}s^2 = \mathbf{F}$$

But we must remember that this is in itself not a complete expression of a physical law: $\mathbf{F}$ is in turn related to the positions of the particles by a further equation, and the entire system is invariant.)

Example Equation (1) is invariant under the transformation (2) with $\mathbf{a} = 0$ and

$$\mathbf{L} = \begin{pmatrix} 1 & 0 & 0 & 0 \\ v^1 & 1 & 0 & 0 \\ v^2 & 0 & 1 & 0 \\ v^3 & 0 & 0 & 1 \end{pmatrix}$$

Proof If we revert to our earlier notation by setting $t = x^0$ and $\mathbf{r} = (x^1, x^2, x^3)$ then (2) becomes, in this special case,

$$\mathbf{r}' = \mathbf{r} + t\mathbf{v}; \quad t' = t$$

where $\mathbf{v} := (v^1, v^2, v^3)$. Equation (2′) is then

$$\mathbf{r} = \mathbf{r}' - t'\mathbf{v} = \mathbf{r}' - t\mathbf{v}$$

Consequently

$$\frac{\mathrm{d}\mathbf{r}}{\mathrm{d}t} = \frac{\mathrm{d}\mathbf{r}'}{\mathrm{d}t} - \mathbf{v} = \frac{\mathrm{d}\mathbf{r}'}{\mathrm{d}t'} - \mathbf{v} \tag{3}$$

and

$$\frac{\mathrm{d}^2\mathbf{r}}{\mathrm{d}t^2} = \frac{\mathrm{d}^2\mathbf{r}'}{\mathrm{d}t^2} = \frac{\mathrm{d}^2\mathbf{r}'}{\mathrm{d}t'^2}$$

Similarly the coordinates $\mathbf{r}_0'$ of the second particle are given by applying (2) to $\mathbf{r}_0$, so that

$$\mathbf{r}_0 - \mathbf{r} = \mathbf{r}_0' - \mathbf{r}'$$

Putting these equations in the equation of motion (1), we find that the effect of the substitution is simply to put dashes on all the coordinates, i.e. (1) is invariant.

The most important mathematical object associated with a physical theory expressed in terms of coordinates is the set of all coordinate transformations under which the laws of the theory are invariant—in special relativity and Newtonian theory, the set of all transformations between inertial coordinates. The sets of transformations we shall be considering here will always form a group, called the *symmetry group* of the theory. (It is plausible that if two successive transformations are performed, each of which leaves the laws unchanged, then the composition of the two also leaves them unchanged. The only complication in verifying that composition really does satisfy the axioms for a group in this case comes when the physical laws involve derivatives, so that it must be shown that everything remains sufficiently differentiable.)

From the laws we can obviously work out the symmetry group. But, conversely, if we know the symmetry group then we also know a great deal about the laws: only those laws are allowed that are invariant under all the transformations of the group, i.e. they are true for all inertial observers in the cases of special relativity and Newtonian theory.

1.3 The Poincaré symmetry group

Special relativity starts from the fact that the equations of Newtonian mechanics and Maxwell's equations for electrodynamics have different symmetry groups. This means that if 'inertial observers' are defined so that the Newtonian equations are true for all of them, then Maxwell's equations will not be true for all of them and the principle of relativity (that the laws of physics are the same for all inertial observers) will be violated.

This is easily seen if we consider the path of a pulse of light in space–time. According to Maxwell's equations, the speed of the pulse should be a fundamental constant, definable in terms of the ratios of certain electromagnetic forces, and so should be the same for all observers for whom Maxwell's equations hold. But two observers in relative motion with coordinates related by the transformation in the example in the previous section, which is a symmetry for Newtonian mechanics, will in general measure different speeds for the pulse—as can be seen from (3) if we take $\mathbf{r}$ to be the position of the pulse at time t. Indeed, if the velocity is a constant in the $\mathbf{r}$ coordinates, then in the $\mathbf{r}'$ coordinates the velocity will be different in different directions.

Experimentally it was found that the speed of light as measured on the Earth was the same in all directions, with no dependence on the different states of motion of the Earth at different points of its orbit round the Sun throughout the year. Relativity theory takes this as meaning that the coordinates that we naturally set up in performing physical measurements are in fact related by transformations that leave Maxwell's equations invariant, and that the true symmetry group of physics is the one that holds for Maxwell's equations, not the Newtonian one. (All the alternatives involve modifying Newtonian theory and/or electromagnetic theory in some complicated and artificial way so as to make them compatible with each other and with observations.)

If this is true, it implies that Newtonian mechanics is wrong and must be replaced by a new set of laws that are invariant under the new symmetry group, called the Poincaré group. We define this group in two steps.

Definition 1 The *Lorentz group* $\mathscr{L}$ is the set of all 4×4 matrices $\mathbf{L}$ satisfying

$$\mathbf{L}^{\mathrm{T}}\mathbf{g}\mathbf{L} = \mathbf{g} \tag{4}$$

where ^{T} denotes transpose and $\mathbf{g}$ is a special matrix given by

$$\mathbf{g} = \begin{pmatrix} -1 & 0 & 0 & 0 \\ 0 & 1 & 0 & 0 \\ 0 & 0 & 1 & 0 \\ 0 & 0 & 0 & 1 \end{pmatrix} \tag{5}$$

Definition 2 The Poincaré group $\mathscr{P}$ is the set of all coordinate transformations of the form (2), i.e.

$$\mathbf{x}' = \mathbf{L}\mathbf{x} + \mathbf{a}$$

with $\mathbf{L}$ a member of $\mathscr{L}$.

It is easily checked that these sets are actually groups: note in particular that from (4) we have $\mathbf{g}\mathbf{L}^{\mathrm{T}}\mathbf{g}\mathbf{L} = \mathbf{g}^2 = \mathbf{I}$ (the unit 4×4 matrix), so that $\mathbf{L}$ is invertible with inverse

$$\mathbf{L}^{-1} = \mathbf{g}\mathbf{L}^{\mathrm{T}}\mathbf{g} \tag{6}$$

a matrix that can be shown to be another element of the Lorentz group.

When O is an inertial observer with coordinates $\mathbf{x}$, we shall idealize O as a point located at the spatial origin; we shall speak of the space–time line $x^1 = x^2 = x^3 = 0$ as the *path* (or *world-line*) of O. If O′ is another inertial observer with coordinates $\mathbf{x}'$, then in his coordinates the path of O, from (2), is again a straight space–time line; in other words, O has constant velocity in the coordinates of O′. Thus it is possible to retain the basic property of Newtonian observers as a physical principle: if O is an inertial observer with inertial coordinates, then O′ is also an inertial observer if and only if O′ is moving with uniform velocity in the coordinates of O. Any two inertial coordinate systems (constructed by inertial observers using some standard procedure) are related by a Poincaré transformation.

1.3.1 Conventions and units

A change in the units of measurement results in a change in the coordinates; changing the unit of length (from metres to feet, say) multiplies the spatial coordinates (x^1, x^2, x^3) by a factor, and changing the unit of time does the same to x^0. Neither of these, nor any combination of them, is compatible with (2) and (4), so that a change of units is not a symmetry of the theory in the rather strict sense in which we have defined it.

One consequence of changing units is that dimensional constants such as the gravitational constant or Planck's constant will alter their numerical values, so that equations in which they appear will not be invariant. However, if we change the units of space and time *by the same factor*, then this numerical change in the fundamental constants will be the only consequence. In particular, the Poincaré group, as defined by (2) and (4), will still be a symmetry transformation within the theory with the new units. However, if we change the units of space and time by different factors, then not only will the fundamental constants change, but the symmetry group of the theory will no longer be given by (4) (the diagonal entries of $\mathbf{g}$ are no longer simply ones).

Because the Poincaré group is so important to the theory, it is natural to choose the relationship between the space and time units in such a way that we do indeed get the simple form (4). Any unbalanced change in units would be as unnatural geometrically as measuring horizontal distances in miles and vertical distances in feet (however useful this may be for certain technical applications). The Poincaré group links the measurement of space and time together: once we have fixed on a unit of distance, the corresponding unit of time is determined. It turns out (see chapter 2) that if distances are measured in cm, then the unit of time is the time taken for light to travel 1 cm.

In these natural units the velocity of light is 1. In books using non-natural units the first component of $\mathbf{g}$ is altered to c^2, where c is the velocity of light, and c appears frequently in formulae.

Another notational variation is that some authors use a $\mathbf{g}$ that is the negative of our $\mathbf{g}$: their equations will often differ from ours in sign. This is purely a matter of convention.

Still another variant is to use x^4 for the time coordinate, resulting in a different order for the components. To establish what convention is being used, the *signature* of the metric is quoted by listing the signs of the diagonal terms in order. Thus our signature is $(-+++)$.

1.3.2 Boosts

A boost is a particular sort of Poincaré transformation in which $\mathbf{a} = 0$ and

$$\mathbf{L} = \mathbf{L}(\alpha) = \begin{pmatrix} \cosh\alpha & -\sinh\alpha & 0 & 0 \\ -\sinh\alpha & \cosh\alpha & 0 & 0 \\ 0 & 0 & 1 & 0 \\ 0 & 0 & 0 & 1 \end{pmatrix}$$

where α is any real number. Its effect is

$$x'^0 = x^0 \cosh\alpha - x^1 \sinh\alpha \tag{7a}$$

$$x'^1 = -x^0 \sinh\alpha + x^1 \cosh\alpha \tag{7b}$$

with the other coordinates unchanged. To be precise, this is 'a boost through the hyperbolic angle α in the (x^0, x^1)-plane'. Boosts in the (x^0, x^2)- and (x^0, x^3)-planes are defined similarly. The reader should verify that (4) is indeed satisfied by $\mathbf{L}(\alpha)$.

Suppose these coordinates are set up by observers O and O′. The spatial point $x'^1 = x'^2 = x'^3 = 0$ in the coordinates of O′, which he regards as his fixed origin, will be a line in space–time that, in the coordinates of O, has the form $x^2 = x^3 = 0$, $x^1 = x^0 \tanh\alpha$. That is, O sees the origin of O′ moving with speed $\tanh\alpha = v$, say. This, then, is a transformation that relates the coordinates of inertial observers who are in relative motion with O′ moving in the x^1-direction of O with speed $v = \tanh\alpha$ and with no relative rotation (so that the x^2 and x^3 coordinates coincide).

In terms of v, (7) becomes

$$x'^0 = (x^0 - vx^1)/\sqrt{(1-v^2)} \tag{8a}$$

$$x'^1 = (x^1 - vx^0)/\sqrt{(1-v^2)} \tag{8b}$$

The converse relationship, expressing $\mathbf{x}$ in terms of $\mathbf{x}'$, is obtained by replacing v by $-v$:

$$x^0 = (x'^0 + vx'^1)/\sqrt{(1-v^2)} \tag{8$'$a}$$

$$x^1 = (x'^1 + vx'^0)/\sqrt{(1-v^2)} \tag{8$'$b}$$

Example (The Lorentz–Fitzgerald contraction.)
Suppose some body occupies the (x^0, x^1)-plane in space–time and that, in the **x**-coordinates, it is stationary and of constant length l. Thus we can take its ends to be the space–time paths $(x^1 = x^2 = x^3 = 0)$ and $(x^1 = l, x^2 = x^3 = 0)$. Figure 1 depicts (shaded) the region occupied by the body.

Another observer O′, using coordinates $\mathbf{x}'$ related to $\mathbf{x}$ by (8), will see the body moving with velocity $-v$. (The $\mathbf{x}'$ coordinates are shown broken in Figure 1.) At the time $x'^0 = 0$ in his coordinates, one end of the body is at the origin, a, and the other is marked by the event b where his x'^1-axis cuts the world-line of the end. Thus the length l' he ascribes to the body is just the x'^1-coordinate of b,

$$l' = x'^1(b)$$

By definition b satisfies

$$x^1(b) = l$$
$$x'^0(b) = 0$$

Thus

$$\begin{aligned} x^0(b) &= vx'^1(b)/\sqrt{(1-v^2)} && \text{from (8'a)} \\ &= vx^1(b) && \text{from (8'b)} \\ &= vl \end{aligned}$$

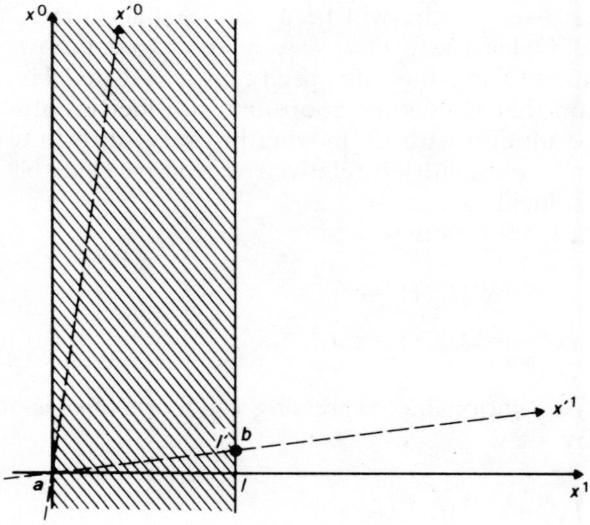

Figure 1

and so

$$l' = x'^1(b) = (x^1(b) - vx^0(b))/\sqrt{(1-v^2)} \qquad \text{from (8b)}$$
$$= l\sqrt{(1-v^2)}$$

Note that l' is shorter than l: in special relativity length becomes a property of the body that varies according to its relation to the observer.

The formulae of this section clearly break down if $v > 1$, when $\sqrt{(1-v^2)}$ becomes imaginary. However, the dynamics of special relativity make it physically impossible to accelerate any object to such a speed. The following argument makes this plausible: take a sequence of observers $O_0, O_1, \ldots$ with coordinates related by boosts, with O_i moving at a speed $v_i = \tanh \alpha_i$ relative to O_{i-1} $(i = 1, 2, \ldots)$. We might expect that in this way we could arrange that the speed of O_i relative to O_0 would increase without bound as $i \to \infty$. However, the transformation relating the coordinates of O_i to those of O_0 is $\mathbf{L}(\alpha_i)\mathbf{L}(\alpha_{i-1}) \ldots \mathbf{L}(\alpha_1) = \mathbf{L}(\Sigma_{r=1}^{i} \alpha_r)$, giving a relative velocity of $\tanh(\Sigma_r \alpha_r) < 1$. Velocity is not additive and is limited by 1 (see also Exercise 4 at the end of this chapter).

1.3.3 Other Lorentz transformations

A *Lorentz transformation* is a Poincaré transformation of the special form $\mathbf{x}' = \mathbf{Lx}$ (i.e. $\mathbf{a} = 0$); a *translation* is one of the form $\mathbf{x}' = \mathbf{x} + \mathbf{a}$: thus any Poincaré transformation is the composition of a Lorentz transformation followed by a translation.

Let $\mathbf{R}$ be a 3×3 matrix satisfying $\mathbf{R}^T\mathbf{R} = \mathbf{I}$, det $\mathbf{R} = 1$; the transformation on Euclidean 3-space defined by such a matrix is called a rotation. Now consider the 4×4 matrix

$$\mathbf{L_R} = \left(\begin{array}{c|ccc} 1 & 0 & 0 & 0 \\ \hline 0 & & & \\ 0 & & \mathbf{R} & \\ 0 & & & \end{array}\right)$$

We can check that (4) is satisfied, so that $\mathbf{L_R}$ is a member of the Lorentz group. We shall call the Lorentz transformation thus defined a *spatial rotation.*

The spatial rotations and boosts together generate a subgroup $\mathscr{L}_+^\uparrow$ of the Lorentz group that is characterized as the set of all Lorentz matrixes $\mathbf{L}$ for which

$$\det \mathbf{L} = 1$$

and $\quad x^0 > 0 \Rightarrow (\mathbf{Lx})^0 > 0$

A general Lorentz transformation can have $\det \mathbf{L} = \pm 1$ (by taking determinants in (4)) and either sign for $(\mathbf{L}\,\mathbf{x})^0$. So to generate the entire Lorentz group we need, in addition to boosts and rotations, two special transformations that violate the preceding conditions for $\mathscr{L}^{\uparrow}_{+}$: they are given by*

$$\mathbf{L}_t = \begin{pmatrix} -1 & 0 & 0 & 0 \\ 0 & 1 & 0 & 0 \\ 0 & 0 & 1 & 0 \\ 0 & 0 & 0 & 1 \end{pmatrix}$$

and $\quad \mathbf{L}_s = -\mathbf{L}_t$

The Lorentz transformations that these define are called time-reflection and space-reflection respectively (for obvious reasons). We shall not, however, be concerned much with transformations of this type. For further details, see Exercise 7 at the end of this chapter.

1.4 Vectors and covectors

We are now going to define vectors and similar objects as geometrical entities which do not depend on one particular coordinate system. We shall find that we can formulate the laws of physics in terms of such geometrical objects alone, which ensures that the laws will be true for all observers and coordinates. If we then select some coordinate system, we can express the vectors in terms of components with respect to these coordinates. On changing the coordinates, these components will change according to precise rules, but the basic geometric and observer-independent nature of the laws will ensure that the laws of physics as a whole are invariant under the coordinate changes of the Poincaré group (in the sense of invariant explained in §1.2.1).

In three-dimensional Euclidean space S, any pair of points (P, Q) defines a vector $\overrightarrow{PQ}$ connecting them. The vector $\overrightarrow{PQ}$ depends only on the difference in coordinates of P and Q, and thus a given vector can be represented in infinitely many ways: P can be chosen arbitrarily and the vector then determines Q.

We now repeat this idea in space–time. We shall associate a vector with each pair of events in the following way: two pairs of events (x, y) and (x', y') will be called *equivalent* (when we write $(x, y) \sim (x', y')$) if

$$\mathbf{x}(y) - \mathbf{x}(x) = \mathbf{x}(y') - \mathbf{x}(x')$$

---

* Though $\mathbf{L}_t$ is the same matrix as $\mathbf{g}$, we use two different names because the matrix appears in quite different roles as $\mathbf{L}_t$ and $\mathbf{g}$.

for some inertial coordinates $\mathbf{x}$, and so for *any* inertial coordinates, because of the linearity of the transformation law (2). The relation $\sim$ is an equivalence relation.

A *vector* is an equivalence class of pairs under this equivalence relation. In other words, we identify a vector as the collection of all the pairs of events that define it.

The vector that is the equivalence class of the pair (x, y) will be written $\overrightarrow{xy}$.

The elements of the column vector $\mathbf{X} := \mathbf{x}(y) - \mathbf{x}(x)$ are called the *components* of the vector $X = \overrightarrow{xy}$ with respect to the coordinates $\mathbf{x}$. (Note that we use corresponding letters for the vector and its components.) If we go to a new coordinate system $\mathbf{x}'$ then it follows immediately from (2) that the components of X with respect to $\mathbf{x}'$ are given by

$$\mathbf{X}' = \mathbf{L}\mathbf{X} \tag{9}$$

Vectors transform by the Lorentz part of a Poincaré transformation.

The set of components in a given coordinate system completely characterizes the vector, so that if we are given such a set in one particular coordinate system, or one set for each coordinate system with the law (9) connecting them, then a unique vector is thereby specified.

For reasons that will emerge in chapter 3, the set of all vectors will be called $T.(M)$. It is a vector space, the operations of addition and multiplication by a scalar being defined in it by considering components. Thus if X and Y are vectors in $T.(M)$ we define the vector $X + Y$ to be the vector whose components in some, and hence in any, coordinate system are $\mathbf{X} + \mathbf{Y}$—and similarly for aX where a is a real number. It is important to realize that, although we have referred to coordinates both to define vectors and to define their addition and scalar-multiplication, these definitions do not depend on *which* coordinates are used, and so are coordinate-independent. This is because the transformation law (9) is linear.

1.4.1 The tangent vector to a curve

This is a particular way of arriving at vectors. In general relativity (§3.2), where the above definition of vectors cannot be used, the idea is taken as the basic definition of a vector.

A *differentiable curve* is a continuous curve γ which maps an interval (a, b) of the real line into space–time M (so that for each parameter value u in (a, b), $\gamma(u)$ is a point of M) and which is such that the functions $x^0(\gamma(u)), \ldots, x^3(\gamma(u))$ that make up $\mathbf{x}(\gamma(u))$ are all differentiable functions of u (for some, and so for any, coordinates $\mathbf{x}$).

The *tangent vector* to a differentiable curve γ at a point $\gamma(u_0)$ on the curve is defined by*

$$\dot{\gamma}(u_0) = \operatorname*{Lim}_{h \to 0} (1/h)(\overrightarrow{\gamma(u_0)\gamma}(u_0+h))$$

Let us write X for $\dot{\gamma}(u_0)$ and consider this expression in terms of a particular coordinate system $\mathbf{x}$. Taking components on both sides gives

$$\begin{aligned}\mathbf{X} &= \operatorname*{Lim}_{h \to 0} (1/h)[\mathbf{x}(\gamma(u_0+h)) - \mathbf{x}(\gamma(u_0))] \\ &= \frac{\mathrm{d}}{\mathrm{d}u}\mathbf{x}(\gamma(u))\bigg|_{u=u_0}\end{aligned}$$

A particular example of a differentiable curve is a coordinate line of some coordinate system, for instance, the x^0-line defined by

$$x^0(\gamma(u)) = u, \qquad x^1(\gamma(u)) = \text{constant},$$
$$x^2(\gamma(u)) = \text{constant}, \quad x^3(\gamma(u)) = \text{constant}.$$

All these coordinate curves have tangent vectors, and the one given is seen, on differentiating by u, to have components

$$\frac{\mathrm{d}\,\mathbf{x}(\gamma(u))}{\mathrm{d}u} = \begin{pmatrix}1\\0\\0\\0\end{pmatrix}$$

If the vector defined in this way is called $\underset{0}{E}$, then applying the same principle to the other coordinates we obtain $\underset{1}{E}$, $\underset{2}{E}$ and $\underset{3}{E}$, with components having a 1 in the position corresponding to the coordinate in question and 0 elsewhere.

These vectors form a basis, in that we can write any other vector as a linear combination of them. This is obvious in components: for any vector X we have

$$\mathbf{X} = \begin{pmatrix}X^0\\X^1\\X^2\\X^3\end{pmatrix} = X^0\begin{pmatrix}1\\0\\0\\0\end{pmatrix} + X^1\begin{pmatrix}0\\1\\0\\0\end{pmatrix} + X^2\begin{pmatrix}0\\0\\1\\0\end{pmatrix} + X^3\begin{pmatrix}0\\0\\0\\1\end{pmatrix}$$

i.e. $$X = X^0\underset{0}{E} + X^1\underset{1}{E} + X^2\underset{2}{E} + X^3\underset{3}{E} \tag{10}$$

* The idea of a limit in $T.(M)$ may be defined by using the natural topology on a finite dimensional vector space, or via coordinates.

1.4.2 Covectors

Vectors in $T.(M)$ are usually called *contravariant vectors*, to distinguish them from a dual sort of vector which we now define.

A *covariant vector* or *covector* is a linear map from $T.(M)$ to the real numbers; the set of covectors is written $T^*_.(M)$.

Thus if ω is a covector, $\omega(X)$ is a real number for all X in $T.(M)$ and

$$\omega(aX + bY) = a\omega(X) + b\omega(Y) \tag{11}$$

for all X and Y in $T.(M)$ and all real a and b. It is well-known from linear algebra that the set of covectors forms a four-dimensional vector space (the *dual space* to $T.(M)$), with addition and real-multiplication being defined by

$$(\lambda + \mu)(X) := \lambda(X) + \mu(X)$$

$$(a\omega)(X) := a\omega(X) \qquad \text{for all } X \text{ in } T.(M)$$

To see this less abstractly we use the basis $\underset{0}{E}, \ldots, \underset{3}{E}$ defined in the previous section. Linearity (i.e. the repeated use of equation (11)) gives, from (10),

$$\omega(X) = X^0\omega(\underset{0}{E}) + X^1\omega(\underset{1}{E}) + X^2\omega(\underset{2}{E}) + X^3\omega(\underset{3}{E})$$

$$= \boldsymbol{\omega}\mathbf{X}$$

where $\boldsymbol{\omega}$ is the *row* matrix $(\omega(\underset{0}{E}), \omega(\underset{1}{E}), \omega(\underset{2}{E}), \omega(\underset{3}{E})) = :(\omega_0, \omega_1, \omega_2, \omega_3)$ and matrix multiplication (to yield a 1×1 matrix, i.e. a number) is implied.

The numbers $\omega_0, \ldots, \omega_3$ are the *components* of ω with respect to the chosen coordinate system, and they characterize the covector uniquely.

If we change coordinates to $\mathbf{x}'$ then $\mathbf{X}$ changes to $\mathbf{X}' = \mathbf{L}\mathbf{x}$, as in equation (9). Suppose that $\boldsymbol{\omega}$ changes to $\boldsymbol{\omega}'$, then we must have

$$\omega(X) = \boldsymbol{\omega}\mathbf{X} = \boldsymbol{\omega}'\mathbf{X}' = \boldsymbol{\omega}'\mathbf{L}\mathbf{X} \qquad \text{for all } X$$

and hence

$$\boldsymbol{\omega} = \boldsymbol{\omega}'\mathbf{L}$$

or

$$\boldsymbol{\omega}' = \boldsymbol{\omega}\mathbf{L}^{-1} \tag{12}$$

1.5 The metric

In the geometry of Euclidean, three-dimensional space a central role is played by the inner product (the dot or scalar product) of two vectors $\mathbf{x}$

and **y**. The analogue to this in relativity is provided by a function g which gives, for the pair of tangent vectors X,Y the number

$$g(X,Y) = \mathbf{Y}^{\mathrm{T}}\mathbf{g}\mathbf{X} = -Y^0X^0 + Y^1X^1 + Y^2X^2 + Y^3X^3 \tag{13}$$

in an inertial coordinate system. If we change to a different coordinate system $\mathbf{x}'$ then we find

$$\mathbf{Y}^{\mathrm{T}}\mathbf{g}\mathbf{X}' = \mathbf{Y}^{\mathrm{T}}\mathbf{L}^{\mathrm{T}}\mathbf{g}\mathbf{L}\mathbf{X} = \mathbf{Y}^{\mathrm{T}}\mathbf{g}\mathbf{X} \qquad \text{from (4)} \tag{14}$$

so that the definition of $g(X,Y)$ does not depend on which coordinate system we use. The function g is called the *metric.*

It is also useful to consider g as a linear map from $T_{\cdot}(M)$ to $T^*_{\cdot}(M)$, in the following way. For each X in $T_{\cdot}(M)$ we define $g(X)$ to be that covector whose value for an arbitrary Y in $T_{\cdot}(M)$ is

$$g(X)(Y) = g(X,Y) \qquad \text{for all } Y \text{ in } T_{\cdot}(M)$$

The metric is the single most basic object in relativity theory. Indeed it would be more logical to start with the metric and then define the Lorentz group as the set of coordinate transformations which leave the metric invariant in the sense of equation (14).

1.5.1 Space, time and the null cone

Just as the metric is analogous to an inner product, so the number

$$g(X,X) = -(X^0)^2 + (X^1)^2 + (X^2)^2 + (X^3)^2 \tag{15}$$

is analogous to the square of the length, $\mathbf{x} \cdot \mathbf{x}$, of a three-dimensional vector. The difference is that (15) can take positive, negative or zero values for non-zero X, allowing us to classify the vectors in $T_{\cdot}(M)$ into three corresponding types. The geometry of this classification is depicted schematically in Figure 2, in which one dimension has been ignored to enable us to make a conventional drawing.

If we fix one point $x_0 \in M$ as origin, then each point y corresponds uniquely to the vector $\overrightarrow{x_0y}$, so the classification of vectors goes over into a classification of points (events) relative to an origin.

Choosing an inertial coordinate system with x_0 as the coordinate origin, then if y lies on the *time*-axis (the x^0-axis) the vector $\overrightarrow{x_0y}$ has coordinates $(y^0,0,0,0)$ and so, from (15), $g(\overrightarrow{x_0y}, \overrightarrow{x_0y})$ is negative. We call all vectors for which (15) is negative *timelike.* It can be shown (Exercise 7 at the end of this chapter) that, given any timelike vector, we can always find an inertial coordinate system in which it has only an x^0-component. Similarly vectors for which $g(X,X)$ is positive are called *spacelike* (and can always be represented as having only spatial components for a suitable coordinate system) while those for which $g(X,X) = 0$ are called *null.* The set of all null vectors is called the *null cone* (see Figure 2).

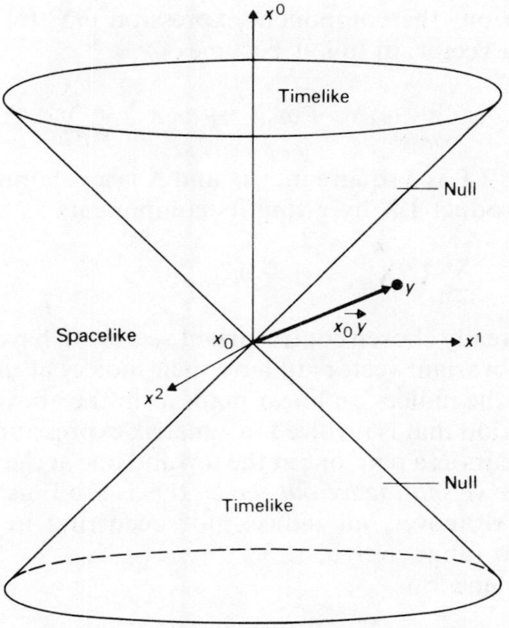

Figure 2. The null cone with vertex at the event x_0. One dimension (x^3) is omitted from the figure.

If two events, x and y, are such that $\overrightarrow{xy}$ is timelike, then the fact that there is a coordinate system in which $\overrightarrow{xy}$ has only a time-coordinate means that there is some observer who sees the two events happening in turn at his spatial origin; in other words, he has enough time to get from one event to the other. But if the spatial separation of the points becomes too great, then $\overrightarrow{xy}$ becomes null and then spacelike, and an observer cannot be present at both successively. Points whose separation is timelike are thus in communication, in the sense that something can get from one to the other; this is not so for points that are spacelike separated. It is an assumption of relativity theory that if x and y are spacelike separated, then nothing happening at x can have any direct causal influence on what happens at y.

1.6 The summation convention

We now introduce a notation that enables us to handle arrays of coordinate-dependent quantities that are more complicated than the vectors and matrices that we have used so far.

If we write out the component-expression $\boldsymbol{\omega}\mathbf{X}$, for the value of a covector on a vector, in full, it becomes

$$\omega X = \omega_0 X^0 + \omega_1 X^1 + \omega_2 X^2 + \omega_3 X^3 = \sum_{i=0}^{3} \omega_i X^i$$

and similarly, if $\mathbf{L}$ is a square matrix and $\mathbf{X}$ is a column matrix, we can specify the product $\mathbf{LX}$ by giving its components as

$$(\mathbf{LX})^i = \sum_{j=0}^{3} \mathbf{L}^i{}_j \mathbf{X}^j \qquad (i = 0, \ldots, 3)$$

We have already chosen contravariant vectors to have their indices at the top and covariant vectors to have their indices at the bottom. If we also position the indices on linear maps as in the above example, then every summation that is involved in a matrix expression will be over an index that occurs as a pair, one at the top and one at the bottom. *In such circumstances we shall leave out the* Σ: this is the Einstein summation convention. Moreover, all indices not occurring in a pair will be understood to range over $0, \ldots, 3$.

Thus from now on

$$\omega_i X^i \quad \text{means} \quad \sum_i \omega_i X^i$$

$$L^i{}_j X^j \quad \text{means} \quad \sum_j L^i{}_j X^j$$

and so on. Every index occurring in a product of factors as a pair, one up and one down, is to be summed over.

To summarize, we now have three notations in use: the direct expression of a quantity in terms of vectors, the matrix expression in terms of the coordinate-components, and the expression with indices using the summation convention also in terms of components.

The close similarity of the notations is brought out if we compare three typical expressions: for a covector on a vector, for the metric on two vectors, and for a trilinear function of three vectors Q. The third goes beyond the capabilities of matrix notation.

	Notation	
geometrical	*matrix*	*index*
$\omega(X)$	$\boldsymbol{\omega}\mathbf{X}$	$\omega_i X^i$
$g(X,Y)$	$\mathbf{Y}^{\mathrm{T}}\mathbf{g}\mathbf{X}$	$Y^i g_{ij} X^j$
$Q(X,Y,Z)$	—	$Q_{ijk} X^i Y^j Z^k$

When the matrix notation can be used it is equivalent to the index notation, and we can translate from matrix to index notation by inserting indices, with the rule that adjacent indices belonging to consecutive matrices that are being multiplied must be summed over, and so must be a pair.

With index notation, unlike matrices, we do not have to worry about the order of the terms, which are simply numbers. Thus the following are equivalent

$$g_{ij}X^iY^j = X^ig_{ij}Y^j = X^iY^jg_{ij}$$

and so on.

Thus index notation in effect uses two sorts of indices: 'free' indices that occur once only in each product of terms and identify to which component of an expression we are referring, and so-called 'dummy' indices, in pairs, which serve to express a summation. Consider, for instance, the expression X^ig_{ij} in the equation

$$\omega_j = X^ig_{ij}$$

On the right hand side we have $\sum_i X^ig_{ij}$ which could also be written $\sum_k X^kg_{kj}$, i.e. X^kg_{kj}. Thus *any other letter can be substituted for a dummy index*. On the other hand X^ig_{il} denotes a different component, and does not match the left-hand-side: *the free indices in all terms must 'match'*. We take the view that only those index-formulae are to be used that correspond to some geometrical expression. Thus an expression like $\omega_i = X^i$ although it is formally meaningful (it could mean $\omega_0 = X^0$, $\omega_1 = X^1$ etc.) will not be allowed because it would be related to the meaningless expression $\omega = X$, relating a covector to a vector. This will be covered if we ensure that free indices also match in their *positions*.

We shall use the summation convention without exception, even if some of the terms are not components. For instance, in § 1.4 we could have written $\sum_i X^i \underset{i}{E}$ (where the X^i are components and the $\underset{i}{E}$ are contravariant vectors) as $X^i\underset{i}{E}$.

1.6.1 Raising and lowering

In relativity we so often use g to map from a contravariant vector X in $T.(M)$ to $g(X)$ in $T^*_\cdot(M)$, and its inverse to map back again, that we come to regard X and $g(X)$ as two different breeds of the same animal, and we reflect this in our notation by writing the components $(g(X))_i = g_{ij}X^j$ simply as X_i. (The position of the index prevents any confusion with the contravariant vector's components X^i.) Conversely, having introduced a covector ω, the expression ω^i will denote the components of the corresponding contravariant vector $g^{-1}(\omega)$, i.e. $\omega^i := (g^{-1})^{ij}\omega_j$.

Raising and lowering are inverse operations: if we raise and then lower an index we return to the original.

Raising and lowering will be applied to all objects with indices. Thus $Q^{i}{}_{jk}$ means $(g^{-1})^{il}Q_{ljk}$, and so on.

1.6.2 Notation for $\mathbf{g}^{-1}$ and I

From the definition of **g** (equation (5)), we see that the inverse matrix is **g** itself ($\mathbf{g}^2 = \mathbf{I}$). Thus the components of the inverse of the geometrical object g are again just **g**. However, g and g^{-1} are quite different geometrically: the first can be regarded as a map from vectors to covectors, the second vice versa; this is reflected in the different position of their indices in the previous subsection. So when **g** is being used as the components of g^{-1} we shall write these components as g^{ij}, with the indices at the top, while g_{ij} will always indicate that we are regarding the (same) components as components of the metric g.

This is consistent with our previous convention for raising and lowering, because according to the raising convention g^{ij} means $(g^{-1})^{ik}(g^{-1})^{jm}g_{km}$. The matrix **g** is *symmetric*, i.e. $(g^{-1})^{jm} = (g^{-1})^{mj}$—it never matters which way round the indices go on either g or g^{-1}. So we have

$$\begin{aligned} g^{ij} &= (g^{-1})^{ik}(g^{-1})^{jm}g_{km} = (g^{-1})^{ik}(g^{-1})^{mj}g_{km} \\ &= ((\mathbf{g}^{-1})\mathbf{g}(\mathbf{g}^{-1}))^{ij} = (g^{-1})^{ij} \end{aligned}$$

Note particularly that this proof does not use the fact that $\mathbf{g}^2 = \mathbf{I}$, and so will still be valid in general relativity when this will not be the case.

From historical tradition, the components of the unit matrix **I** are written δ^i_j (a symbol called the Kronecker delta). Thus

$$\delta^i_j \begin{cases} = 1 & \text{if } i = j \\ = 0 & \text{otherwise.} \end{cases}$$

The defining equation $\mathbf{IX} = \mathbf{X}$ becomes $\delta^i_j X^j = X^i$; more generally we have equations like $\delta^i_j Q_{ikl} = Q_{jkl}$. Thus multiplying by δ^i_j and summing over j (or i) results in the *replacement* of the index j by the index i, or vice versa.

Warning. $\delta^i_j = 1$ when $i = j$, but $\delta^i_i = \sum_i \delta^i_i = 4$ (not 1!).

1.6.3 More on transformation laws

Proposition A Lorentz transformation matrix **L** satisfies

$$(\mathbf{L}^{-1})^i{}_j = L_j{}^i \tag{16}$$

Proof 1 By the raising/lowering conventions,

$$\begin{aligned} L_j^{\,i} := g_{jk}g^{il}L^k{}_l = g^{il}L^{\mathrm{T}}{}_l{}^k g_{jk} &= g^{il}L^{\mathrm{T}}{}_l{}^k g_{kj} \qquad \text{(by symmetry of } g) \\ &= (\mathbf{g}\mathbf{L}^{\mathrm{T}}\mathbf{g})^i{}_j = (\mathbf{L}^{-1})^i{}_j \end{aligned}$$

from equation (6). q.e.d.

It is a useful exercise to prove this from the basic definitions, without using equation (6).

Proof 2 The definition of a Lorentz transformation matrix is equation (4), which, in index notation, is

$$L^i{}_j g_{ik} L^k{}_l = g_{jl} \tag{17}$$

Multiplying by g^{lm} and summing over l gives

$$L^i{}_j g_{ik} L^k{}_l g^{lm} = g_{jl} g^{lm} = \delta^m_j$$

(since $\mathbf{g}\mathbf{g}^{-1} = \mathbf{I}$). Multiplying by $(L^{-1})^j{}_n$ and summing over j gives

$$\begin{aligned}((L^{-1})^j{}_n L^i{}_j)(g_{ik} L^k{}_l g^{lm}) &= \delta^m_j (L^{-1})^j{}_n = (L^{-1})^m{}_n \\ &= (\delta^i_n)(L_i{}^m) = L_n{}^m\end{aligned}$$

as required.

If we now apply this result to the transformation of covectors, given by equation (12), we obtain the following.

Corollary Under the transformation L, the covector ω transforms according to

$$\omega'_j = L_i{}^j \omega_j \tag{18}$$

We can compare this with the corresponding equation for the transformation of vectors (9), which in index notation becomes

$$X'^i = L^i{}_j X^j \tag{19}$$

1.6.4 Compulsory exercise

(for readers who are not familiar with the summation convention). All the objects appearing in this exercise may be regarded as matrices or row/column vectors. The main rule to observe is that throughout any manipulation indices should be the same *only* if they are to be summed over.

Example Eliminate Z from the equations

$$X^i = P^i{}_j Z^j \qquad \text{(i)}$$

$$Z_i = R_{ij} W^j \qquad \text{(ii)}$$

Raising the index i on both sides of (ii) gives

$$Z^k = R^k{}_j W^j \qquad \text{(iii)}$$

We cannot immediately replace k by j and substitute for Z^j in (i), because that would give $P^i{}_j R^j{}_j W^j$ where it is not clear what summation has to be done. So first relabel (iii) by changing j to l and k to j, giving

$$Z^j = R^j{}_l W^l$$

and now substitute in (i) to give

$$X^i = P^i{}_j R^j{}_l W^l$$

After a while this sort of relabelling should become automatic.

1 Given that

$$\mathbf{X} = [X^i] = \begin{pmatrix} 1 \\ 0 \\ 0 \\ 1 \end{pmatrix}, \ \mathbf{Y} = [Y^i] = \begin{pmatrix} 0 \\ 1 \\ 1 \\ 1 \end{pmatrix}, \ \mathbf{P} = [P^i{}_j] = \begin{pmatrix} 1 & 0 & 1 & 0 \\ 0 & 0 & 0 & 0 \\ 2 & 0 & 3 & 1 \\ 0 & 0 & -1 & 0 \end{pmatrix}$$

decide whether the following objects are numbers, columns (vectors), rows (covectors) or matrices, and work them out explicitly.
(i) $X_i X^i$ (ii) $Y_i Y^i$ (iii) $X_i Y^i$ (iv) $X^i Y^j$ (v) $P_{ij} X^j$ (vi) $P^i{}_i$
(vii) $P_{ij} - P_{ji}$ (viii) $P_{ij} + P_{ji}$.

2 Prove the following:
(i) if $X^i Y_i = Z^i Y_i$ for all Y, then $X = Z$ (Hint: consider $[Y_i] = (1, 0, 0, 0)$, then $[Y_i] = (0, 1, 0, 0)$ etc.);
(ii) if $P^i{}_j Y^j = 0$ for all $\mathbf{Y}$, then $P^i{}_j = 0$;
(iii) if $P^i{}_j Y^j = Y^i$ for all Y, then $C^i{}_j = 0$, where $C^i{}_j = P^i{}_j - \delta^i_j$.

3 Show that if $\mathbf{R}$ is a non-singular matrix with coefficients $R_i{}^j$, its inverse having coefficients $(R^{-1})_i{}^j$, then

$$R_i{}^j (R^{-1})_{jk} = g_{ik}$$

4 A matrix $\mathbf{P} = [P^i{}_j]$ is such that, for all $\mathbf{X}$

$$P^i{}_j Z^j = X^i, \quad \text{where } Z_i = P_{ji} X^j$$

Using the foregoing example and exercises, show that $\mathbf{P}$ is a member of the Lorentz group.

5 (i) Show that, if $P_{ij} X^i Y^j = 0$ for all X and Y, then $P_{ij} = 0$.
(ii) Find a non-zero matrix $\mathbf{P}$ such that $\mathrm{P}_{ij} \mathrm{X}^i \mathrm{X}^j = 0$ for all X.
(iii) Show that, if $P_{ij} = P_{ji}$ and $P_{ij} X^i X^j = 0$ for all X, then $P_{ij} = 0$.
(iv) Show that, if $P_{ij} = P_{ji}$ and $S_{ij} = -S_{ji}$, then $P_{ij} S^{ij} = 0$.

6 Given any matrix $\mathbf{R}$, find a real number a such that $R_{ij} = C_{ij} + a g_{ij}$ for some matrix $\mathbf{C}$ satisfying $C_{ij} g^{ij} = 0$.

Further exercises will be found at the end of the chapter.

1.7 Tensors

We have already encountered covectors as linear maps of vectors into the real numbers, and the metric $g(X, Y)$ as a bilinear map (i.e. linear in

X and linear in Y) into the real numbers. We can obviously generalize the idea to a multilinear map $Q(X, Y, \ldots, Z) = Q_{ij\cdots m}X^iY^j \ldots Z^m$ from n vectors into the real numbers, linear in each argument. And we could generalize even further, to a multilinear map of s vectors and r covectors, of the form $F(X, Y, \ldots, Z, \lambda, \mu, \ldots, \omega) = F_{ij\cdots m}{}^{pq\cdots t}X^iY^j \ldots Z^m \lambda_p\mu_q \ldots \omega_t$ into the real numbers. Such a map is called a *tensor of type* (r, s). Note that r is the number of upper indices and s is the number of lower indices on the components of the tensor. (The proof that any such multilinear map can be specified by a set of components is just a repetition of the argument of §1.4.2.)

Other examples that we have already met are as follows:

(i) a vector is a tensor of type (1,0) and its value on a covector ω can be defined to be $X^i\omega_i (= \omega(X))$;
(ii) the unit matrix δ^i_j is a tensor of type (1, 1) and its value on a vector X and covector ω is $\delta^i_jX^j\omega_i (= \omega(X))$;
(iii) the inverse metric g^{-1} is a tensor of type (2, 0).

The set of all tensors of type (m, n) is written $T^{(m,n)}_{\cdot}(M)$.

We recall that we also regarded the metric tensor, of type (0, 2), as a linear map from $T_{\cdot}(M)$ to $T^*_{\cdot}(M)$, by writing $g(X, Y)$ as $g(X)(Y)$. In same way, any tensor can be considered in different ways, for instance:

Proposition Any multilinear function $F:(T^*_{\cdot}(M))^r \times (T_{\cdot}(M))^s \to T^{(r',s')}_{\cdot}(M)$, taking r covectors and s vectors into a tensor of type (r', s'), can be regarded as a tensor of type $(r+r', s+s')$, and vice versa.

Proof Given such an F, we simply specify its value on the $(r+r')$ covectors $\overset{1}{\omega}, \overset{2}{\omega}, \ldots, \overset{r+r'}{\omega}$ and the $(s+s')$ vectors $\underset{1}{X}, \underset{2}{X}, \ldots, \underset{s+s'}{X}$ to be the number

$$F(\underset{1}{X}, \underset{2}{X}, \ldots, \underset{s}{X}, \overset{1}{\omega}, \overset{2}{\omega}, \ldots, \overset{r}{\omega})(\underset{s+1}{X}, \underset{s+2}{X}, \ldots, \underset{s+s'}{X}, \overset{r+1}{\omega}, \overset{r+2}{\omega}, \ldots, \overset{r+r'}{\omega})$$

1.7.1 Products of tensors

Suppose U is a tensor of type (r, s) and V is one of type (r', s'). Then we can combine them to give a tensor of type $(r+r', s+s')$, called the tensor-product of U and V, written $U \otimes V$. In terms of components, the definition is that we multiply the components of U and the components of V:

$$(U \otimes V)_{ij\cdots lm\cdots p}{}^{qr\cdots tu\cdots x} = U_{ij\cdots l}{}^{qr\cdots t}V_{m\cdots p}{}^{u\cdots x}$$

In terms of linear maps, this is equivalent to

$$(U \otimes V)(\underset{1}{X}, \ldots, \underset{s}{X}, \underset{s+1}{X}, \ldots, \underset{s+s'}{X}, \overset{1}{\omega}, \ldots, \overset{r}{\omega}, \overset{r+1}{\omega}, \ldots, \overset{r+r'}{\omega})$$

$$= U(\underset{1}{X}, \ldots, \underset{s}{X}, \overset{1}{\omega}, \ldots, \overset{r}{\omega}).\ V(\underset{s+1}{X}, \ldots, \underset{s+s'}{X}, \overset{r+1}{\omega}, \ldots, \overset{r+r'}{\omega})$$

For example, the tensor-product of two vectors, X and Y, is a tensor of type (2, 0) with components $X^i Y^j$.

1.7.2 Transformation laws

The transformation law for the tensor $X \otimes Y$ follows immediately from the law (19) for a vector:

$$(X \otimes Y)'^{ij} = L^i{}_k L^j{}_l (X \otimes Y)^{kl}$$

and we can derive a similar law from (18) for the tensor-product of a covector with a vector or another covector. These are all special cases of the general tensor transformation law

$$T'_{ij\cdots l}{}^{pq\cdots t} = L_i{}^m L_j{}^n \ldots L_l{}^o L^p{}_u L^q{}_v \ldots L^t{}_w T_{mn\cdots o}{}^{uv\cdots w}$$

where we simply have one L for each index, with the obvious summations. The proof of this follows from evaluating $T(X, \ldots, Z, \omega, \ldots, v)$ in x-coordinates and in x'-coordinates, exactly as in the derivation of the law (12) for a covector.

1.7.3 Contraction

A tensor of type (1, 1), with components $T_i{}^j$, gives rise to a number

$$\mathrm{Tr}(T) := T_i^i$$

called its *trace*. This definition is independent of the coordinate system, because in a different coordinate system we have

$$T'^i_i = L_i{}^k L^i{}_l T_k{}^l = \delta^k_l T_k{}^l \qquad \text{from (16)}$$

$$= \mathrm{Tr}(T)$$

Similarly, if we are given a tensor T of type $(r+1, s+1)$, then contraction gives a tensor of type (r, s) as follows. By the proposition of § 1.7, we can regard T as a map which takes r covectors $\omega, \mu, \ldots$ and s vectors $X, Y, \ldots$ into a tensor of type (1, 1); if we take the trace of this, we get a map from r covectors and s vectors into a number. In terms of coordinates

$$(X, Y, \ldots, Z, \omega, \mu, \ldots, v) \xrightarrow{T} T_{ab\cdots hi}{}^{pq\cdots tj} X^a Y^b \ldots Z^h \omega_p \mu_q \ldots \omega_t \xrightarrow{\mathrm{Tr}}$$

$$\rightarrow T_{ab\cdots hi}{}^{pq\cdots ti} X^a Y^b \ldots Z^h \omega_p \mu_q \ldots \omega_t$$

The map is multilinear and so defines a tensor of type (r, s), with components $T_{ab\cdots hi}{}^{pq\cdots ti}$, called the contraction of T (with respect to the two indices i that are summed over). We can similarly define contractions $T_{a\cdots i\cdots k}{}^{q\cdots i\cdots l}$ over any other pair of indices.

1.7.4 Position of indices

From the tensor with components $T_{i\cdots k}{}^{p\cdots q\cdots s}$ we may define a tensor with components $T_{i\cdots k}{}^{p\cdots}{}_{q}{}^{\cdots s}$ by the 'lowering' convention as $T_{i\cdots k}{}^{p\cdots}{}_{q}{}^{\cdots s} = g_{ql} T_{i\cdots k}{}^{p\cdots l\cdots s}$. Note that in this case we maintain the order of the indices, and do not 'tidy up' the tensor by writing all the lower indices first, as we have done up to now. The order must be preserved so as to record which index it is that has been lowered.

1.8 Paths and particles

From now on by a curve we shall always mean a differentiable curve in the sense of § 1.4.1; moreover, we shall assume that all our curves have the tangent vector $\dot{\gamma}$ continuous (there are no sharp corners) and non-zero (the parameter does not 'stand still'). For simplicity of writing we introduce the notation

$$\mathbf{x}(\gamma(u)) =: \gamma(u) = \begin{pmatrix} \gamma^0(u) \\ \gamma^1(u) \\ \gamma^2(u) \\ \gamma^3(u) \end{pmatrix}$$

so that the tangent vector has components $\dot{\gamma}^i(u) = (\mathrm{d}/\mathrm{d}u)\gamma^i(u)$.

A *particle* is always thought of as an object idealized to a point in three-dimensional space, and so to a path in space–time. This path is called the *world-line* of the particle. We assume that the path is given by a curve $\gamma(u)$, and that at any point $\gamma(u_0)$ on this path we could in principle have an observer O_0 who *at that event* was instantaneously precisely keeping pace with the particle. For him there would, at that instant, be no change in the spatial position of the particle, so that in his coordinates $\dot{\gamma}^1(u_0) = \dot{\gamma}^2(u_0) = \dot{\gamma}^3(u_0) = 0$ (and consequently $\dot{\gamma}^0(u_0) \neq 0$, since we are assuming that $\dot{\gamma} \neq 0$). This means that $\dot{\gamma}$ is *timelike*, at any point of the path.

1.8.1 Reparametrization

The path of a particle, in the sense of the set of points in space–time, does not uniquely fix the curve, that is the map, that we use to describe it: we could reparametrize the path by using instead the map

$\gamma(\theta(s)) = :\tilde{\gamma}(s)$, where θ is any smooth, strictly increasing real function. The tangent vectors in the two parametrizations are related by

$$\dot{\tilde{\gamma}}(s_0) = \dot{\gamma}(\theta(s_0))\left.\frac{\mathrm{d}\theta}{\mathrm{d}s}\right|_{s_0} \tag{20}$$

(from the chain rule for differentiating a function of a function). We now choose θ so that, in the coordinates of the inertial observer O_0 referred to above (with $u_0 = \theta(s_0)$) we have $\dot{\tilde{\gamma}}^0(s_0) = 1$, by taking $\mathrm{d}\theta(s_0)/\mathrm{d}s = 1/\dot{\gamma}^0(u_0)$. Since $\tilde{\gamma}^0$ is the time coordinate of the particle, the condition $(\mathrm{d}\tilde{\gamma}^0/\mathrm{d}s) = 1$ means that we are parametrizing by the time-coordinate, as measured by the observer who is instantaneously keeping pace with the particle. Indeed, we can set up this parametrization along the whole of the curve, not just at one point, by solving the differential equation $\mathrm{d}\theta/\mathrm{d}s = 1/\dot{\gamma}^0(\theta(s))$ along the curve. The solution is unique up to an added constant. The parameter $s = \theta^{-1}(u)$ defined in this way is called the *proper time* of the particle.

By definition, in the coordinates of O_0 we have $\dot{\tilde{\boldsymbol{\gamma}}}^{\mathrm{T}} = (1, 0, 0, 0)$ and so $g(\dot{\tilde{\gamma}}, \dot{\tilde{\gamma}}) = -1$ (independently of coordinates). Thus using (20) and the bilinearity of g we have

$$g(\dot{\tilde{\gamma}}, \dot{\tilde{\gamma}}) = (d\theta/ds)^2 g(\dot{\gamma}, \dot{\gamma}) = -1 \tag{21}$$

This gives the convenient equation for θ (without reference to a special observer)

$$\mathrm{d}\theta/\mathrm{d}s = [-g(\dot{\gamma}, \dot{\gamma})]^{-\frac{1}{2}}$$

In physical terms, a time-coordinate is something that is measured by a clock, and the time-coordinate of an inertial coordinate system should correspond to the time kept by an (ideal) clock carried by an inertial observer. The instants in the life of an inertial observer are labelled by the times indicated on such a clock, so that the clock defines a parameter t, say, on his world-line, a parameter that coincides with the coordinate x^0. (x^0 is defined everywhere, whereas the parameter t just labels the points on the world-line.)

Imagine now a non-inertial observer with path $\gamma(s)$, undergoing accelerations, and carrying another clock which labels the points on his world-line by the parameter s. This parameter will depend on how the clock is affected by the accelerations: pendulum clocks and atomic clocks, for instance, would not be expected to agree when they are being accelerated. Let us define a *perfect clock* to be one that is not affected at all by acceleration, in the sense that at any moment, with parameter s_0, it runs at the same rate as an inertial clock carried by the inertial observer who at that instant is momentarily keeping pace with (i.e. has zero velocity relative to, and is coincident with) the accelerating observer. 'Runs at the same rate' means that the clock-parameters s and t ($t = x^0$) of the two observers satisfy

$$\left.\frac{dx^0(\gamma(s))}{ds}\right|_{s=s_0} = 1$$

which is precisely the definition of s as a proper time. Thus *a perfect clock measures proper time.*

Note that in formulating this we have used the coordinate time x^0 (defined everywhere) rather than the parameter time t (defined only on the world-line of the inertial observer): in fact the equation $dt(\gamma(s))/ds = 1$ would not make sense because t will in general not be defined at $\gamma(s)$ unless the two observers are actually coincident for an extended period. The distinction between t and x^0 is essential, because t is measured directly by a clock whereas x^0 is a mathematically constructed coordinate. The parameter s is also measured by a (perfect) clock; in chapter 3 we shall consider the possibility of constructing a coordinate to correspond to it (Fermi–Walker coordinates).

If we compare s with the coordinate time of an inertial observer O′ who is *not* keeping pace with $\gamma(s)$ at some instant, then we find that the coordinate time x'^0 and the proper time s satisfy

$$\frac{dx'^0(\gamma(s))}{ds} > 1$$

(see Exercise 5 and equation (2) of chapter 2). This is confirmed, with the expected numerical agreement, by the observation that unstable particles take longer to decay when they are moving. If we imagine that their decay is governed by an 'internal clock', which measures s, then the laboratory time x'^0 that elapses during an interval from s_1 to s_2 of internal (proper) time is *greater* than $s_2 - s_1$ if the particles are moving relative to O′, the laboratory observer:

$$x'^0(\gamma(s_2)) - x'^0(\gamma(s_1)) = \int_{s_1}^{s_2} \frac{dx'^0(\gamma(s))}{ds} ds > s_2 - s_1$$

1.8.2 Particle kinetics

The idea touched on in this subsection is developed at length in chapter 2.

If the particle path is described by a curve $\gamma(s)$ parametrized by proper time (we now omit the $\tilde{}$) then the vector $\dot{\gamma}(s)$ is called the 4-*velocity* of the particle at the event $\gamma(s)$.

Every physical particle has a constant number m associated with it, called its *mass*, and the covector* $mg(\dot{\gamma}(s))$ is called the 4-*momentum* at $\gamma(s)$, with components $m\dot{\gamma}_i(s)$.

An essential part of Newton's laws of motion is expressed in the two following relativistic laws.

Law 1 The 4-momentum of a free particle is constant on its path. (This implies that the mass is constant, as already stated, and that the particle moves in a straight line.)

Law 2 If N free particles come together and interact in any way so as to produce N' free particles afterwards, the sum of the 4-momenta of the particles before the interaction is equal to the sum of the 4-momenta afterwards.

This law (of which Law 1 is a special case with $N = N' = 1$) is the law of *conservation of* 4-*momentum* for special relativity.

1.9 Differentiation

1.9.1 The differential

The differential of a differentiable function is a special, but important, case of a covector (analogous to the tangent vector to a curve in the case of contravariant vectors). The differential then generalizes to tensors, giving a useful means of generating further tensors.

A real valued function f on space–time M can either be specified directly, or, if we take some coordinate system $\mathbf{x}$, we can give the value $\tilde{f}(\mathbf{x})$ of the function at the point with coordinates $\mathbf{x}$, thus

$$\tilde{f}(\mathbf{x}(x)) = f(x)$$

The same symbol is frequently used for both $\tilde{f}$ and f, but they are really functions of different variables.

The function f is called *differentiable* at an event $x \in M$ if $\tilde{f}$ is differentiable at $\mathbf{x}(x)$ (for some, and hence any, coordinate system). In this case the differential of f at x is defined to be the covector whose components are

$$\nabla f_i(x) \quad (\text{or } \mathrm{d}f_i(x)) := \partial \tilde{f}/\partial x^i|_{\mathbf{x}(x)}$$

Both notations ∇f and $\mathrm{d}f$ are used for this covector, because there are two separate operations d and ∇ which happen to coincide when applied to ordinary functions.

We must check that this always gives the same covector, whatever

*Momentum is defined as a covector rather than a vector because in quantum theory and Hamiltonian dynamics momentum appears naturally as a 'dual' of position or velocity.

coordinates are used. In a second system $\mathbf{x}'$, the chain rule for partial derivatives gives

$$\frac{\partial \tilde{f}}{\partial x^i} = \frac{\partial x'^j}{\partial x^i}\frac{\partial \tilde{f}}{\partial x'^j} \qquad \text{(summation over } j) \tag{22}$$

Now $x'^j = L^j{}_k x^k$, and so $\partial x'^j/\partial x^i = L^j{}_k \partial x^k/\partial x^i = L^j{}_k \delta^k_i$, since $\partial x^k/\partial x^i$ is zero unless $k = i$, in which case it is 1. Thus $\partial x'^j/\partial x^i = L^j{}_i$ and so (22) gives

$$\partial\tilde{f}/\partial x^i = L^j{}_i \, \partial\tilde{f}/\partial x'^j$$

This is the transformation law (18) for a covector, as required for consistency; the $\mathbf{x}'$ coordinates define the same covector.

In particular, we can apply this to the coordinate functions $x^0(x)$, $x^1(x)$ etc. and form the covectors dx^0, dx^1, dx^2, and dx^3. These have components $(dx^i)_j = \partial x^i/\partial x^j = \delta^i_j$, and so they form a basis for $T^*_\cdot(M)$, just as the vectors $\underset{i}{E}$ did for $T_\cdot(M)$. If ω is an arbitrary covector, then

$$\omega = \omega_i \, dx^i$$

(which can be proved by taking components of both sides). In particular, applying this to df, where f is a function on M, we have

$$df = \frac{\partial \tilde{f}}{\partial x^i} dx^i$$

just as in ordinary calculus.

1.9.2 Application to tensors

We now consider a tensor-valued function on space–time M, i.e. at each event x there is specified a tensor $T(x)$. Such a function is called a *tensor field* (or a vector or covector field if the tensor is of type $(1, 0)$ or $(0, 1)$ respectively). We shall only consider tensor fields where the tensor is everywhere of the same type. As with functions, the field is called *differentiable* if the corresponding functions in terms of coordinates

$$T_{ij\cdots l}{}^{pq\cdots t}(\mathbf{x}(x)) := (T(x))_{ij\cdots l}{}^{pq\cdots t}$$

are differentiable functions of the coordinates. If T is of type (m, n), then its differential is defined to be the tensor field of type $(m, n+1)$ whose components are

$$(\nabla T(x))_{ij\cdots kl}{}^{pq\cdots t} = \frac{\partial T_{ij\cdots k}{}^{pq\cdots t}}{\partial x^l}$$

$$=: T_{ij\cdots k}{}^{pq\cdots t}{}_{,l}$$

From now on we shall always denote partial derivatives with respect to the coordinates in this way, by a comma.

The verification that this is independent of the choice of coordinates is the same as in the previous subsection.

Of course, we can keep applying this operation to produce tensors of steadily increasing type. We shall denote the components of these higher derivatives by

$$(\nabla\nabla \ldots \nabla T)_{ij\cdots klm\cdots o}{}^{pq\cdots t} = T_{ij\cdots k}{}^{pq\cdots t}{}_{,lm\cdots o}$$
$$= \frac{\partial^n T_{ij\cdots k}{}^{pq\cdots t}}{\partial x^l \, \partial x^m \ldots \partial x^o}$$

i.e. we only put in one comma.

1.9.3 Directional derivatives

We are frequently interested in the rate at which the value of a function (or a tensor) varies in moving along the path of a particle, i.e. the derivative

$$\dot{f} := \frac{\mathrm{d}}{\mathrm{d}s} f(\gamma(s))$$

In terms of coordinates, using the coordinate version $\tilde{f}$ of f, we have from the chain rule

$$\dot{f} = \frac{\mathrm{d}}{\mathrm{d}s} \tilde{f}(\mathbf{x}(\gamma(s))) = \frac{\partial \tilde{f}}{\partial x^i} \frac{\mathrm{d}x^i(\gamma(s))}{\mathrm{d}s}$$
$$= (\nabla f)(\dot{\gamma})$$

It is useful to have a special notation for this combination, so we define

$$\underset{X}{\nabla} f := (\nabla f)(X) = X^i f_{,i} \tag{23}$$

Thus if $X = \dot{\gamma}$ is the tangent vector to a curve, $\underset{X}{\nabla}$ is the operator that differentiates f along the curve. It is called the *directional derivative* in the direction X.

We can extend its definition to tensors by defining

$$(\underset{X}{\nabla} T)_{ij\ldots l}{}^{pq\cdots t} = X^k T_{ij\ldots l}{}^{pq\cdots t}{}_{,k}$$

Since this is the contraction of the tensor product of two tensors it is again a tensor.

Exercises

1 A map $h\colon = T.(M) \to T.(M)$ is defined by $h(X) = X + g(U,X)U$ where $U \in T.(M)$ is a fixed vector with $g(U,U) = -1$.
(i) Give an expression for the components $h^i{}_j$ of h regarded as a tensor of type (1, 1) (viz. $(h(X))^i = h^i{}_j X^j$).
(ii) Prove that $h^2 = h$ (i.e. $h^i{}_j h^j{}_k = h^i{}_k$). Interpret h geometrically.

2 Show that two vectors X and Y are proportional if and only if $X^i Y^j - Y^i X^j = 0$.

3 Verify that the boost defined by (7) is a Lorentz transformation.

4 Verify the relation $\mathbf{L}(\alpha)\mathbf{L}(\beta) = \mathbf{L}(\alpha+\beta)$ (§1.3.2). If three inertial frames $\mathbf{x}$, $\mathbf{x}'$ and $\mathbf{x}''$ are related by boosts in the (x^0, x^1)-planes, with the velocity of the spatial origin of $\mathbf{x}''$ relative to $\mathbf{x}'$ being v_2 and that of $\mathbf{x}'$ relative to $\mathbf{x}$ being v_1, show that the velocity of the spatial origin of $\mathbf{x}''$ relative to $\mathbf{x}$ is $(v_1 + v_2)/(1 + v_1 v_2)$.

5 Two events occur at space–time points x and y having coordinates $(0, 0, 0, 0)$ and $(\tau, 0, 0, 0)$ respectively, in a given inertial system. Show that the time interval between x and y (the difference in the time coordinates) in another coordinate system related to the first by a boost with velocity v is greater than τ and is given by $\tau/\sqrt{(1-v^2)}$. (This effect is known as the *dilatation of time*.)

6 Verify that the axioms for a group are satisfied by the set of all Lorentz matrices, defined by (4).

7 Let X be a given timelike vector.
(i) Show that there exists a spatial rotation $\mathbf{R}$ such that $(\mathbf{RX})^2 = (\mathbf{RX})^3 = 0$.
(ii) Find a boost in the (x^0, x^1)-plane $\mathbf{L}(\alpha)$ such that

$$(\mathbf{L}(\alpha)\mathbf{RX}) = \begin{pmatrix} t \\ 0 \\ 0 \\ 0 \end{pmatrix}$$

where $-t^2 = g(X,X)$.
(iii) Deduce that if γ is a particle path (a straight line with timelike tangent vector), then there exists a Lorentz transformation to coordinates $\mathbf{x}'$ such that $x'^1(\gamma(s)) = x'^2(\gamma(s)) = x'^3(\gamma(s)) = 0$, i.e. the particle is at rest at the spatial origin.

8 Show that the sets

$$\mathscr{L}^{\uparrow} := \{\mathbf{L}\in\mathscr{L} : \mathrm{L}^{0}{}_{0} > 0\}$$

$$\mathscr{L}_{+} := \{\mathbf{L}\in\mathscr{L} : \det \mathbf{L} > 0\}$$

are subgroups of $\mathscr{L}$.

9 **L** is any Lorentz transformation and the vectors $\underset{i}{\mathbf{E}}$ are the usual basis vectors with components $\underset{i}{E}{}^{j} = \delta^{j}_{i}$. $\mathbf{L}(\alpha)$ and **R** are chosen as in Exercise 7 so that $\mathbf{L}(\alpha)\mathbf{R}(\mathbf{L}\underset{0}{\mathbf{E}}) = (\pm 1, 0, 0, 0)$. Define $\underset{i}{\mathbf{E}}' := \mathbf{L}(\alpha)\mathbf{R}\mathbf{L}\underset{i}{\mathbf{E}}$.
(i) By considering $g(\underset{i}{E}, \underset{j}{E})$ show that $\underset{\alpha}{E}'^{0} = 0$ and that the numbers $\underset{\alpha}{E}'^{\beta}$ $(\alpha, \beta = 1, 2, 3)$ are the components of a 3×3 orthogonal matrix **S** (i.e. $\mathbf{S}^{\mathrm{T}}\mathbf{S} = \mathbf{I}$).
(ii) Deduce that there is a spatial rotation matrix **T** such that either $\mathbf{T}\mathbf{L}(\alpha)\mathbf{R}\mathbf{L} = \pm\mathbf{I}$ or $\mathbf{T}\mathbf{L}(\alpha)\mathbf{R}\mathbf{L} = \pm\mathbf{L}_{t}$ (§1.3.3).
(iii) Show that any Lorentz matrix **L** can be written in the form $\mathbf{Q}\mathbf{R}^{-1}\mathbf{L}(\alpha)\mathbf{S}$ where **R** and **S** are spatial rotations, $\mathbf{L}(\alpha)$ is a boost in the (x^0, x^1)-plane and

$\mathbf{Q} = \mathbf{I}$ if $\mathbf{L}\in\mathscr{L}^{\uparrow}_{+}$

$\mathbf{Q} = -\mathbf{I}$ if $\det \mathbf{L} = 1$ and $L^{0}{}_{0} < 0$

$\mathbf{Q} = \mathbf{L}$ if $\det \mathbf{L} = -1$ and $L^{0}{}_{0} < 0$

$\mathbf{Q} = -\mathbf{L}$ if $\det \mathbf{L} = -1$ and $L^{0}{}_{0} > 0$

2
Further special relativity

Although the material in this chapter is not absolutely necessary for general relativity, many readers will probably find it helpful to read at least the first few sections. The main aim of the chapter is to develop the mathematics behind the tensor T that will enter the equations of general relativity. A simplified version of T is given in §2.2.2, with a fuller version incorporating electromagnetism in §2.3.4. Only the briefest outlines of the topics are given: the reader who wants a deeper exposition of what is going on is urged to consult a text devoted to special relativity.

2.1 Particle collisions and decays

The commonest application of special relativity is to the beams of very fast particles produced by accelerators. When these strike particles in a 'target' which are almost at rest relative to the laboratory, various particles are emitted, again at high velocity; some are unstable and subsequently decay into two or more further particles. For a full description, quantum mechanics should be used, but the fundamental ideas that motivate the quantum mechanical description are based on the simple model of point particles given in §1.8.

2.1.1 More kinetic formulae

To compare with laboratory measurements we must express the momentum P in terms of the 3-velocity $\mathbf{v} = (\mathrm{d}x^1/\mathrm{d}t, \mathrm{d}x^2/\mathrm{d}t, \mathrm{d}x^3/\mathrm{d}t)$, where we revert to t for x^0.

When the world-line of the particle is parametrized by proper time its tangent vector, in the notation of §1.8, has components

$$\dot{\gamma}^i = \frac{\mathrm{d}x^i(\gamma(s))}{\mathrm{d}s} = \frac{\mathrm{d}x^i}{\mathrm{d}t}\frac{\mathrm{d}t}{\mathrm{d}s}$$

Since $\mathrm{d}x^0/\mathrm{d}t \equiv 1$, this gives us

$$\dot{\boldsymbol{\gamma}} = \frac{\mathrm{d}t}{\mathrm{d}s}\begin{pmatrix} 1 \\ \mathrm{d}x^1/\mathrm{d}t \\ \mathrm{d}x^2/\mathrm{d}t \\ \mathrm{d}x^3/\mathrm{d}t \end{pmatrix}$$

and so the corresponding covariant vector has components (lowering the index as in §1.6.1) which form the row matrix

$$\dot{\boldsymbol{\gamma}}^{\mathrm{T}}\mathbf{g} = \frac{\mathrm{d}t}{\mathrm{d}s}(-1, \mathbf{v}) \tag{1}$$

The condition for parametrization by proper time (cf. equation (21) of chapter 1) gives

$$\begin{aligned} -1 &= g(\dot{\gamma}, \dot{\gamma}) \\ &= \dot{\boldsymbol{\gamma}}^{\mathrm{T}}\mathbf{g}\dot{\boldsymbol{\gamma}} = \left(\frac{\mathrm{d}t}{\mathrm{d}s}\right)^2(-1+v^2) \end{aligned}$$

where $v^2 := \mathbf{v}\,\mathbf{v}^{\mathrm{T}}$ is the square of the speed, and so

$$\frac{\mathrm{d}t}{\mathrm{d}s} = (1-v^2)^{-\frac{1}{2}} =: \beta(v) \tag{2}$$

Thus from (1) and (2) the components of the momentum are given by

$$\mathbf{P} = m\dot{\boldsymbol{\gamma}}^{\mathrm{T}}\mathbf{g} = m\beta(v)(-1, \mathbf{v}) \tag{3}$$

P satisfies the equation

$$g^{-1}(P, P) = g(m\dot{\gamma}, m\dot{\gamma}) = -m^2 \tag{4}$$

2.1.2 Mass and energy

The last three components of P form the 3-*momentum* $\mathbf{p} = m\beta(v)\mathbf{v}$, which, as part of P, is conserved according to Law 2 of §1.8.2. If we define $m^* := m\beta(v)$, then $\mathbf{p} = m^*\mathbf{v}$, which is an equation that has the same form as the Newtonian expression for the momentum. Because of this Newtonian correspondence, m^* is called the *apparent mass*†.

The first component P^0 can be expanded, using (3) and (2), as

$$-P^0 = m\beta(v) = m + \tfrac{1}{2}mv^2 + \mathrm{O}(v^2)$$

(for small v). This can also be expressed as

$$-P^0 \approx \text{mass} + \text{kinetic energy}$$

This already suggests that mass and energy might be of the same nature. But they are actually interconvertible, as the following consideration shows.

Imagine two equal inelastic masses (of clay, for instance) with momenta $m\beta(-1, \mathbf{v})$ and $m\beta(-1, -\mathbf{v})$ colliding from opposite directions and sticking together into a single lump with momentum $M(-1, \mathbf{0})$. Conservation of momentum applied to P^0 gives

†Some books use m for the apparent mass, denoting our 'mass' by m_0 which is then called the 'rest mass'.

$M = 2m\beta \approx 2(m+\frac{1}{2}mv^2)$: in other words, the kinetic energies of the two separate masses have been incorporated into an additional mass mv^2 for the combination.

Accordingly, we may regard mass and kinetic energy both as types of energy, and identify $E := -P^0 = m^*$ as the *total energy* of the particle. In non-natural units where time is measured in seconds and distance in metres, this relation becomes $E = m^*c^2$. In our units $c = 1$.

To summarize

$$\mathbf{P} = (-m^*, m^*\mathbf{v}) = (-\text{energy}, \text{3-momentum})$$

where $m^* = \beta m$.

2.1.3 Particle decays

We shall consider a single particle of mass M, moving on a timelike straight line (according to Law 1 of §1.8.2), which decays into two fragments. It is possible (see chapter 1, Exercise 7) to choose coordinates so that the world-line of the initial particle is specified by $x^1 = x^2 = x^3 = 0$, i.e. we can adopt an inertial frame of reference in which the particle is at rest at the spatial origin. Afterwards it splits into two fragments with velocities $\mathbf{v}_1$ and $\mathbf{v}_2$, and masses m_1 and m_2 respectively.

Using (3), the initial momentum and the two final momenta then have the following components:

$$\mathbf{P} = M\beta(0)(-1, \mathbf{0}) = (-M, \mathbf{0})$$

$$\mathbf{p}_1 = m_1\beta(\mathrm{v}_1)(-1, \mathbf{v}_1) = (-E_1, \mathbf{p}_1)$$

$$\mathbf{p}_2 = m_2\beta(\mathrm{v}_2)(-1, \mathbf{v}_2) = (-E_2, \mathbf{p}_2)$$

Now applying equation (4) gives

$$g^{-1}(p_i, p_i) = (-E_i^2 + \mathbf{p}_i^2) = -m_i^2 \qquad (i = 1, 2) \tag{5}$$

while conservation of momentum (Law 2 of §1.8.2) gives

$$\mathbf{P} = \mathbf{p}_1 + \mathbf{p}_2 \Rightarrow M = E_1 + E_2 \tag{6}$$

$$\mathbf{0} = \mathbf{p}_1 + \mathbf{p}_2 \tag{7}$$

We solve for the energies E_1 and E_2. Equation (7) gives $p_1 = -p_2$, which from (5) gives

$$E_1^2 - m_1^2 = E_2^2 - m_2^2$$

Eliminating E_2 between this equation and (6) now gives

$$E_1 = \frac{M^2 - m_2^2 + m_1^2}{2M} \tag{8}$$

Similarly

$$E_2 = \frac{M^2 - m_1{}^2 + m_2{}^2}{2M}$$

For each fragment the velocity is then determined from (8) and (3), viz.

$$\frac{m_i}{\sqrt{(1 - v_i{}^2)}} = E_i \qquad (i = 1, 2)$$

Note that for this to be possible we need $E_1 \geq m_1$, which from (8) is equivalent to $M \geq m_1 + m_2$, i.e. the combined mass–energy of the fragments must be less than that of the original with the difference appearing as kinetic energy. This process is, of course, simply the reverse of our considerations concerning the lumps of clay.

2.1.4 Particle collisions

This is extensively covered in texts on special relativity. The basic technique is to use once again coordinates in which the total momentum of the system has zero spatial components. Two incident particles would, in these coordinates, have momenta $(-E_1, \mathbf{p})$ and $(-E_2, -\mathbf{p})$. There is a total energy of $E_1 + E_2$ available for the production of one, two, or more particles emerging from the collision. Once their masses are specified, the law of conservation of momentum places constraints on their possible velocities, and hence on their energies.

In practice one is interested also in coordinates determined by the laboratory, in which the target particle (number 2) is at rest, and it is interesting to compute the different value of the energy, E'_1, of the incident particle number 1, in these coordinates.

Supposing the particles to have equal mass m we have

$$E_1 = E_2 = \beta m, \qquad \mathbf{p} = m\beta\mathbf{v}$$

To transform to the laboratory frame we must perform a boost with 3-velocity $-\mathbf{v}$, so as to reduce the 3-velocity of particle 2 to zero. If we align our coordinates so that $\mathbf{v} = (v, 0, 0)$ then the incident particle 1 has momentum $(-E, \mathbf{p}) = (-\beta m, m\beta v, 0, 0)$; applying the standard boost formula equation (8) of chapter 1 with v replaced by $-v$ to this covector gives for its momentum in the laboratory frame

$$(-E'_1, p') = \beta(-\beta m - m\beta v^2, 2m\beta v, 0, 0)$$

and thus

$$E'_1 = \frac{m(1+v^2)}{1-v^2}$$

$$= \frac{2E_1{}^2}{m} - m$$

(eliminating v in terms of E_1).

So we see that when E_1 is large the energy E'_1 in the laboratory frame that must be given to the moving particle increases as the square of the energy E_1 that is actually available for the production of new particles—an effect that makes the process of reaching higher energies progressively more difficult. It can be overcome by allowing beams of particles to collide head-on, when all the energy becomes available.

2.2 Densities and continua

2.2.1 The flux vector

To pass from particles to a continuum we shall imagine a stream of particles: a closely spaced family of timelike curves in M (each supposed parametrized by proper time s). We fix attention on a point $x_0 = \gamma(u_0)$ on one of these curves. As we described in §1.8, we can use an inertial coordinate system $\mathbf{x}'$ with origin at x_0 and the time axis tangent to the curve, so that $\gamma(u_0)'^i = 0$, $\dot{\gamma}(u_0)'^i = \delta^i_0$. An observer who uses this frame will at time $x'^0 = 0$ see the particle momentarily at rest at his spatial origin.

At this instant he will find a spatial distribution of the other particles near the origin, which he will describe by a density n'^0, i.e. the number of particles in a volume V' (small compared with the length-scale of changes in n'^0 but large enough for us to ignore the 'grain' of the picture) being $N' = n'^0 V'$.

We shall suppose that all the particles in V' have approximately the same velocities, of which the vector $X = \dot{\gamma}(u_0)$ is typical. In the continuum picture this typical local velocity X varies smoothly with position, and at x_0 we have $X'^i = \delta^i_0$.

Now compare this with the viewpoint of an observer in uniform motion relative to the first, but whose coordinates $\mathbf{x}$ have the same origin in space–time. By making spatial rotations of their coordinates the two observers can ensure that their direction of motion is along their x^1-axes, with their x^2- and x^3-axes coinciding.

Define a volume V in the hyperplane $x^0 = 0$ of this observer to consist of all points which have $x^0 = 0$ and the triple (x'^1, x'^2, x'^3) in V'. This means that V is the projection of V' onto the hyperplane $\{x^0 = 0\}$ along the world-lines of the particles (see Figure 3).

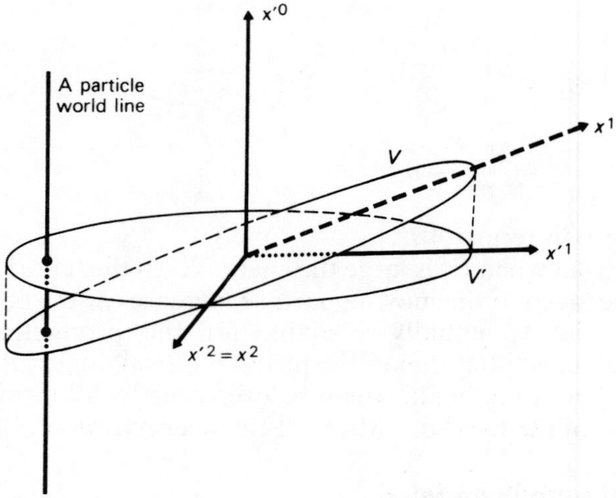

Figure 3. Volumes V and V'at $x^0=0$. One dimension ($x^3,=x^{3'}$) is not shown, so that the volumes appear as surfaces.

Comparing the situation with that of the Lorentz contraction (Figure 1, p. 8) we see that the dimensions of V' and V in the x'^1 and x^1 directions are related by the same formulae as for the Lorentz contraction, but with **x** and **x**′ interchanged; the dimensions of V and V' in the x^2 and x^3 directions are the same. Thus the magnitudes of the two volumes are related by

$$V = V'\sqrt{(1-v^2)}$$

By construction, the particles in V are precisely those that are in V', so that the **x**-observer sees the same number $N = N'$ of particles, with a density

$$n^0 = N/V = (n'^0 V')/(V'\sqrt{(1-v^2)}) = n'^0\beta \tag{9}$$

These particles, in the **x**-coordinates, are all moving with a 3-velocity $\mathbf{v} = (-v, 0, 0)$. So if we define a column matrix **n** by

$$\mathbf{n} = \begin{pmatrix}\text{Number density}\\ \text{Flux 3-vector}\end{pmatrix} = \begin{pmatrix}\mathrm{n}^0\\ \mathrm{n}^0\mathbf{v}^{\mathrm{T}}\end{pmatrix}$$

then from (9)

$$\mathbf{n} = n'^0\beta\begin{pmatrix}1\\ \mathbf{v}^{\mathrm{T}}\end{pmatrix} = \mathrm{n}'^0\mathbf{X}$$

i.e. n is proportional to the representative velocity of the particles X. Thus **n** comprises the components of a contravariant vector $n = n'^0X$, called the flux vector.

2.2.2 The energy–momentum tensor

Now we assume that each particle has mass m, and hence momentum $P = mg(X)$. Define

$$T := P \otimes n$$

with components

$$T_i^0 = P_i n^0 = (-(\text{mass-density}), \text{3-momentum density})$$
$$T_i^\alpha = P_i n^0 v^\alpha = (\text{flux of } P_i)^\alpha \qquad (\alpha = 1, 2, 3)$$

The fact that the mass-density $-T_0{}^0$ appears as one component of a tensor, and is not a scalar as in the Newtonian theory, is crucial for the general relativistic theory of gravitation. It suggests that the 'potential' of the gravitational field, linked to the mass-density, must also be a tensor.

Note that

$$T^{ij} = n'^0 m X^i X^j, \qquad \text{and so}$$
$$T^{ij} = T^{ji} \tag{10}$$

If we have k different streams of particles, each will define an energy–momentum tensor. In this case we define* the total energy–momentum tensor as

$$T := T_{(1)} + T_{(2)} + \dots + T_{(k)}$$

Since momentum, energy, and their fluxes are additive the physical interpretation of this T remains the same; for example, the 3-vector (T_i^1, T_i^2, T_i^3) is the total flux of P_i, made up of contributions from different particles carrying momentum in different directions.

2.2.3 The conservation law

We fix on a spatial volume V with boundary S. Then we can write

$$\int_V \frac{\partial}{\partial x^0} T_i^0 \, \mathrm{d}V = \frac{\mathrm{d}}{\mathrm{d}x^0} \int_V T_i^0 \mathrm{d}V = \frac{\mathrm{d}}{\mathrm{d}x^0} \int_V n^0 P_i \, \mathrm{d}V$$
$$= (\text{rate of change of momentum in } V)_i$$
$$= -(\text{net flux of momentum out of } V)_i$$

(since, by conservation of momentum, the momentum in V can change only by being carried out by particles crossing the boundary S)

$$= -\int_S \sum_{\alpha=1}^{3} n^0 P_i v^\alpha \mathrm{d}S_\alpha = -\int_S \sum_{\alpha=1}^{3} T_i^\alpha \mathrm{d}S_\alpha$$

$$= \int_V \left(-\sum_{\alpha=1}^{3} (T_i^\alpha)_{,\alpha} \right) \mathrm{d}V$$

---

* This additivity of the energy–momentum tensor only holds for the simple 'kinetic' picture we are considering here; it does not hold for the T due to fields, for instance.

by Gauss's theorem (the divergence theorem). Since V is arbitrary we can equate the integrands to get

$$\frac{\partial}{\partial x^0} T_i^0 = -\sum_{\alpha=1}^{3} (T_i^\alpha)_{,\alpha}$$

or

$$T_i{}^j{}_{,j} = 0 \tag{11}$$

Note that the conservation of momentum that we have used to derive this expression holds even if the particles are colliding with each other. Thus we can extend the validity of (11) to the kinetic theory of gases, in which, rather than a fixed number k of separate streams of particles we have a continuous range of possible directions, and the total T is expressed as an integral over the range of particle directions, rather than as a sum. Despite this formal alteration, the argument would go through unchanged.

The equation (11) is in effect the form taken by the law of conservation of energy and momentum for a continuum, and its validity is a fundamental physical principle.

2.2.4 Macroscopic interpretation

In the previous subsection we had the equation

$$(\text{rate of change of momentum in } V)_i = -\int_S \sum_{\alpha=1}^{3} T_i^\alpha \mathrm{d}S_\alpha$$

which is a result that still holds true when electromagnetic and other fields are included in T. In Newtonian theory the rate of change of 3-momentum is simply the force, so in this terminology the spatial components of this equation are

$$(\text{force on } V)_\beta = -\int_S \sum_\alpha T_\beta{}^\alpha \mathrm{d}S_\alpha$$

Thus it is reasonable to interpret $-\sum T_\beta{}^\alpha \mathrm{d}S_\alpha$ as the force on the surface element $\mathrm{d}S_\alpha$ which is precisely the definition of the 3×3 matrix $[T_\beta{}^\alpha]$ as the Newtonian stress-tensor. If we combine this with the interpretation of T_i^0 in §2.2.2 and use the symmetry $T_{ij} = T_{ji}$ (10), we have

$$[T_{ij}] = \left[\begin{array}{c|c} \text{mass–energy density } (=\rho) & \text{3-momentum density} \\ \hline \underbrace{\text{3-momentum density}}_{\text{1 column}} & \underbrace{\text{stress}}_{\text{3 columns}} \end{array}\right] \begin{array}{l} \}\,\text{1 row} \\ \}\,\text{3 rows} \end{array}$$

If T is diagonal (and in general we can make a coordinate transformation so that it is) then the momentum density is zero (meaning that we are in the coordinates of an observer moving with the matter so that there is no net relative momentum of the matter around him) and the diagonal components, T_{11}, T_{22} and T_{33} are just the direction-dependent *pressures* p_1, p_2 and p_3, p_α being the inward force per unit area on a surface normal to the x^α-axis. Thus in this special case

$$[T_{ij}] = \begin{bmatrix} \rho & 0 & 0 & 0 \\ 0 & p_1 & 0 & 0 \\ 0 & 0 & p_2 & 0 \\ 0 & 0 & 0 & p_3 \end{bmatrix}$$

2.3 Electromagnetism

A moving charged particle generates an electromagnetic field round it that is characterized by two 3-vector fields $\mathbf{E}$ and $\mathbf{B}$. The field exerts a force $e(\mathbf{E}+\mathbf{v}\wedge\mathbf{B})$ on any other particle of charge e moving with 3-velocity $\mathbf{v}$. Thus in Newtonian theory the motion of this second particle will be given by

$$\frac{\mathrm{d}^2\mathbf{x}}{\mathrm{d}t^2} = \frac{e}{m}(\mathbf{E}+\mathbf{v}\wedge\mathbf{B}) \tag{12}$$

This does not have the required invariance under the Lorentz group, and so is not acceptable in special relativity: we look for a relativistic expression, written in terms of geometrical quantities, to which equation (12) is an approximation, valid for low velocities.

Since $\mathrm{d}s/\mathrm{d}t \approx 1$ for slow particles, the left-hand-side of (12) is very nearly equal to the spatial part of the vector $\mathrm{d}^2\mathbf{x}/\mathrm{d}s^2$, suggesting that we should replace it by this vector in forming the relativistic law.

On the right-hand-side the term $\mathbf{v}\wedge\mathbf{B}$ poses a problem as the vector product is an essentially three-dimensional construction. To solve this we first rewrite $\mathbf{v}\wedge\mathbf{B}$ as

$$\mathbf{v}\wedge\mathbf{B} = \begin{pmatrix} 0 & B_z & -B_y \\ -B_z & 0 & B_x \\ B_y & -B_x & 0 \end{pmatrix}\begin{pmatrix} v_x \\ v_y \\ v_z \end{pmatrix} = \mathbf{F}\mathbf{v} \approx \mathbf{F}\frac{\mathrm{d}\mathbf{x}}{\mathrm{d}s} \tag{13}$$

thus defining a 3×3 matrix $\mathbf{F}$. Now the extension to four dimensions is performed by extending $\mathbf{F}$ to a 4×4 matrix (the components of a tensor), incorporating the term involving $\mathbf{E}$ in the process. Thus we look for a law of the form

$$\frac{\mathrm{d}^2\mathbf{x}}{\mathrm{d}s^2} = \frac{e}{m}\mathbf{F}\frac{\mathrm{d}\mathbf{x}}{\mathrm{d}s} \tag{14}$$

where

$$\mathbf{F} = \begin{pmatrix} F^0{}_0 & F^0{}_1 & F^0{}_2 & F^0{}_3 \\ F^1{}_0 & 0 & B_z & -B_y \\ F^2{}_0 & -B_z & 0 & B_x \\ F^3{}_0 & B_y & -B_x & 0 \end{pmatrix} = \left(\begin{array}{c|c} a & \mathbf{f} \\ \hline \mathbf{f}' & \mathbf{F} \end{array}\right)$$

which corresponds to a tensor with components $F^i{}_j$.

Separating the space and time parts of the summation on the right-hand-side of (14) gives

$$\frac{\mathrm{d}^2x^i}{\mathrm{d}s^2} = \frac{e}{m}F^i{}_j\frac{\mathrm{d}x^j}{\mathrm{d}s} = \frac{e}{m}\left(\sum_{\alpha=1}^{3} F^i{}_\alpha\frac{\mathrm{d}x^\alpha}{\mathrm{d}s} + F^i{}_0\frac{\mathrm{d}x^0}{\mathrm{d}s}\right) \tag{15}$$

We want this to approximate to (12) when we are dealing with a slowly moving particle, for which $\mathrm{d}x^0/\mathrm{d}s \approx 1$; in this case the components $i = 1, 2, 3$ of (15) give the 3-vector equation

$$\frac{\mathrm{d}^2\mathbf{x}}{\mathrm{d}x^2} = \frac{e}{m}\left(\mathbf{F}\frac{\mathrm{d}\mathbf{x}}{\mathrm{d}s} + \mathbf{f}'\right)$$

from which we see that we obtain agreement with (12) if $\mathbf{f}' = \mathbf{E}$, i.e. $(F^1{}_0, F^2{}_0, F^3{}_0) = (E_x, E_y, E_z)$.

The remaining components of F are determined by the requirement that s is a proper time parameter, i.e.

$$-1 = g(\dot{\gamma}, \dot{\gamma}) = g_{ij}\frac{\mathrm{d}x^i}{\mathrm{d}s}\frac{\mathrm{d}x^j}{\mathrm{d}s}$$

Differentiating gives

$$0 = g_{ij}\left(\frac{\mathrm{d}^2x^i}{\mathrm{d}s^2}\frac{\mathrm{d}x^j}{\mathrm{d}s} + \frac{\mathrm{d}x^i}{\mathrm{d}s}\frac{\mathrm{d}^2x^j}{\mathrm{d}s^2}\right) = 2g_{ij}\frac{\mathrm{d}^2x^i}{\mathrm{d}s^2}\frac{\mathrm{d}x^i}{\mathrm{d}s} \quad \text{(symmetry of } g\text{)}$$

$$= 2g_{ij}F^i{}_k\frac{\mathrm{d}x^k}{\mathrm{d}s}\frac{\mathrm{d}x^j}{\mathrm{d}s} = F_{kj}\frac{\mathrm{d}x^k}{\mathrm{d}s}\frac{\mathrm{d}x^j}{\mathrm{d}s}$$

For this to be true for all $\mathrm{d}x^i/\mathrm{d}s$ we must have $F_{kj} = -F_{jk}$. Thus $\mathbf{f}' = \mathbf{f}$ and $a = 0$. Thus the type $(1, 1)$ tensor F has components

$$\mathbf{F} = \begin{pmatrix} 0 & E_x & E_y & E_z \\ E_x & 0 & B_z & -B_y \\ E_y & -B_z & 0 & B_x \\ E_z & B_y & -B_x & 0 \end{pmatrix} \tag{16}$$

2.3.1 Transformation properties

Let us apply a boost in the (x^0, x^1)-plane with velocity $v \ll 1$, so that $\beta(v) \approx 1$. Then $F^2{}_3$, $F^2{}_2$, $F^3{}_2$ and $F^3{}_3$ are unchanged, and because of $F_{ij} = -F_{ji}$ the diagonal terms remain zero. For the remaining terms

$$\begin{aligned} E'_x = F'^0{}_1 &= L^0{}_0L_1{}^1F^0{}_1 + L^0{}_1L_1{}^1F^1{}_1 + L^0{}_0L_1{}^0F^0{}_0 + L^0{}_1L_1{}^0F^1{}_0 \\ &\approx E_x \end{aligned}$$

(the terms in $F^1{}_1$ and $F^0{}_0$ being zero, the other terms being $O(v^2)$). Similarly

$$E'_y = F'^0{}_2 = L^0{}_0L_2{}^2F^0{}_2 + L^0{}_1L_2{}^2F^1{}_2 \approx E_y + vB_z$$

and so on, leading to

$$\mathbf{E}' \approx \mathbf{E}+\mathbf{B}\wedge\mathbf{v}, \qquad \mathbf{B}' \approx \mathbf{B}-\mathbf{E}\wedge\mathbf{v} \tag{17}$$

where $\mathbf{v} = (v, 0, 0)$.

The first equation of (17) is Faraday's law of induction, which states that a frame moving in a magnetic field $\mathbf{B}$ will see an extra electric field of magnitude $\mathbf{B}\wedge\mathbf{v}$ (the basis of the dynamo). The second equation shows how a moving charge can produce its magnetic field: boosting the pure electric field of a stationary charge gives rise to a magnetic field of $-\mathbf{E}\wedge\mathbf{v}$.

It is striking that these two basic connections between electric and magnetic fields arise from the transformation law of a single tensor.

2.3.2 Sources

A stationary charge e situated at the point with spatial coordinates $\mathbf{r}_0$ produces a pure electric field $\mathbf{E}(r) = e(\mathbf{r}-\mathbf{r}_0)/|\mathbf{r}-\mathbf{r}_0|^3$. It is well known that this can be expressed by saying that if we have a collection of N charges, each of strength e, in a volume V bounded by a surface S, then

$$\int_S \mathbf{E}\cdot\mathbf{dS} = 4\pi eN \tag{18}$$

and

$$\nabla\wedge\mathbf{E} = \mathbf{0} \tag{19}$$

Now we pass to a continuum model by considering a stream of particles moving in parallel with a common velocity. Choosing coordinates as in § 2.2 with respect to which the particles are at rest we can write (18) as

$$\int_S \mathbf{E}'\cdot\mathbf{dS} = 4\pi e\int_V n'^0 \mathrm{d}V$$

whence Gauss's theorem (the divergence theorem) gives

$$\nabla\cdot\mathbf{E}' = 4\pi e n'^0$$

or (since we have $\mathbf{B} = 0$ at the moment and $n'^\alpha = 0$ *for* $\alpha = 1, 2, 3$)

$$F'^{ij}{}_{,j} = 4\pi e n'^i = : J'^i \tag{20}$$

Although derived in a particular coordinate system, all the items in the equation are tensors and so we can transform to any other coordinate system, in which the equation will still be valid.

Putting $i = 0$ in (20), and using the form (16) for F gives

$$\nabla\cdot\mathbf{E} = 4\pi e n'^0$$

while the remaining three components give

$$\nabla \wedge \mathbf{B} = 4\pi e n'^{0}\mathbf{v} + \mathrm{d}\mathbf{E}/\mathrm{d}t$$

Thus (20) is equivalent to two of Maxwell's equations. We assume it to be true for any stream of particles (whether parallel or not), which is equivalent to regarding the value of $F^{i}{}_{j}$ as depending only on the particles at the point in question, the motion of others further away being irrelevant.

To rewrite (19) in relativistic form requires more technical devices. As with the reformulation of $\mathbf{v} \wedge \mathbf{B}$, we express the vector product by forming a matrix with the vector concerned in the off-diagonal entries. This time, in order to express $\nabla \wedge \mathbf{E}$ we need to form a matrix with $\mathbf{E}$, rather than $\mathbf{B}$, in the spatial part of the matrix. To do this we first define a tensor ε to have the following components in some inertial frame:

$$\varepsilon_{ijkl} = \begin{cases} 1 & \text{if the values of } (ijkl) \text{ form an even permutation of } (0123) \\ -1 & \text{for an odd permutation} \\ 0 & \text{otherwise} \end{cases}$$

If we transform to another frame the new components are

$$\varepsilon'_{ijkl} = L_i{}^m L_j{}^n L_k{}^q L_l{}^p \varepsilon_{mnqp} \tag{21}$$

Now if $(ijkl) = (0123)$ the right-hand-side of this becomes simply the definition of the determinant det $\mathbf{L}$. Then, if we interchange two indices, say i and j, we can write

$$\varepsilon'_{jikl} = L_i{}^n L_j{}^m L_k{}^q L_l{}^p \varepsilon_{mnqp} = L_i{}^m L_j{}^n L_k{}^q L_l{}^p \varepsilon_{nmqp} = -\varepsilon'_{ijkl}$$

It is easy to see that these properties imply that (21) takes the form

$$\begin{aligned} \varepsilon'_{ijkl} &= (\det \mathbf{L})\, \varepsilon_{ijkl} \\ &= \varepsilon_{ijkl} \end{aligned}$$

providing that we restrict ourselves to $\mathscr{L}_{+}^{\uparrow}$ (see § 1.3.3). Thus, subject to this restriction, ε is a tensor that can be defined independently of a choice of coordinates.

Now set

$${}^{*}F_{ij} := \tfrac{1}{2}\varepsilon_{ijkl}F^{kl}$$

This will be a tensor, satisfying ${}^{*}F_{ij} = -{}^{*}F_{ji}$ (antisymmetry), like F_{ij}. Its components are easily calculated, for example

$$\begin{aligned} {}^{*}F_{01} &= \tfrac{1}{2}(\varepsilon_{0123}F^{23} + \varepsilon_{0123}F^{32}) \quad \text{(other terms zero)} \\ &= F^{23} \qquad \text{(from the antisymmetry of both } F \text{ and } \varepsilon) \\ &= B_x \end{aligned}$$

In this way we find

$$(*F^i{}_j) = \begin{pmatrix} 0 & -B_x & -B_y & -B_z \\ -B_x & 0 & E_z & -E_y \\ -B_y & -E_z & 0 & E_x \\ -B_z & E_y & -E_x & 0 \end{pmatrix}$$

i.e. $*F$ is related to F by replacing $\mathbf{E}$ by $-\mathbf{B}$ and $\mathbf{B}$ by $\mathbf{E}$. It is called the *dual* of F.

Returning at last to our recasting of equation (19), we find that in a frame where $\mathbf{B} = \mathbf{0}$ it becomes

$$*F^{ij}{}_{,j} = 0 \tag{22}$$

This equation has the same content as (20), but with $\mathbf{E}$ and $\mathbf{B}$ interchanged as described and no J-term; it thus corresponds to the other two Maxwell's equations

$$\nabla \cdot \mathbf{B} = 0$$

$$\nabla \wedge \mathbf{E} = -\mathrm{d}\mathbf{B}/\mathrm{d}t$$

We can express (22) in terms of F as follows. With $i = 1$ (22) gives

$$0 = *F_{1j,}{}^{j} = \tfrac{1}{2}\varepsilon_{1jkl}F^{kl,j}$$

$$= \varepsilon_{1230}F^{30,2} + \varepsilon_{1302}F^{02,3} + \varepsilon_{1023}F^{23,0}$$

(using $F^{30,2} = -F^{03,2}$ etc.)

$$= F^{30,2} + F^{02,3} + F^{23,0}$$

Similar equations hold for the other values of i and so we conclude that, if i, j and k are distinct,

$$F^{ij,k} + F^{jk,i} + F^{ki,j} = 0 \tag{23}$$

However, if two of i, j, k are the same, $i = j = 1$ for instance, then the left-hand-side of (23) takes the form

$$F^{11,k} + F^{1k,1} + F^{k1,1}$$

which is zero because $F^{11} = 0$ and $F^{1k} = -F^{k1}$. Thus (23) holds for all values of the indices.

2.3.3 Electromagnetic waves

We now study the basic equations (20) and (22) in the special case where there are no charges present, so that

$$F^{ij}{}_{,j} = *F^{ij}{}_{,j} = 0$$

Consider an F of the form

$$F_{ij} = A_{i,j} \tag{24}$$

for some vector field A. Then

$$*F_{ij,}{}^{j} = \varepsilon_{ijkl}A^{k,lj} \tag{25}$$

Now $A^{k}{}_{,lj}$ is symmetric in (l, j), while ε_{ijkl} is antisymmetric in these indices. We saw in § 1.6.4, Exercise 5 (iv) that the contraction of a symmetric and an antisymmetric index-pair yields zero, thus (25) shows that (22) is identically satisfied.

Conversely, it is possible to show that any solution of (22) can be written in the form (24), though this is more difficult; indeed, for any such F there is a wide choice of possible A-fields, and we can use this scope to require in addition that

$$A^{i}{}_{,i} = 0 \tag{26}$$

(noting that the values of $A_{i,j}$ with $i = j$ that are involved here do not affect the value of F as given by (24)).

Now substitute (24) in (20) to get

$$\begin{aligned} 0 = F^{ij}{}_{,j} &= A^{i,j}{}_{j} - A^{j,i}{}_{j} \\ &= A^{i,j}{}_{j} - A^{j}{}_{,j}{}^{i} \\ &= A^{i,j}{}_{j} \qquad \text{from (26)} \end{aligned}$$

Using the explicit form of the metric, this is

$$\left(-\frac{\partial^2}{\partial x^{0\,2}} + \frac{\partial^2}{\partial x^{1\,2}} + \frac{\partial^2}{\partial x^{2\,2}} + \frac{\partial^2}{\partial x^{3\,2}}\right)A^{i} = 0 \tag{27}$$

which is the familiar wave equation.

A particular set of solutions consists of fields of the form

$$A^{i} = C^{i}\mathscr{R}e\,\exp(\mathrm{i}k_{l}x^{l}) \tag{28}$$

which represents a plane wave moving in the direction of the vector k. This will satisfy (27) provided that C and k are constant with k a null vector ($k_i k^i = 0$). To check this, note that

$$(k_l x^l)_{,j} = k_l(x^{l}{}_{,j}) = k_l \delta^{l}_{j} = k_j$$

so that

$$A^{i}{}_{,j} = C^{i}\mathscr{R}e\,\mathrm{i}k_j \exp(\mathrm{i}k_l x^l) \tag{29}$$

and

$$A^{i}{}_{,j}{}^{j} = C^{i}\mathscr{R}e(-k_j k^j)\exp(\mathrm{i}k_l x^l)$$

Also, from (29) we see that (26) is satisfied provided $C_l k^l = 0$.

2.3.4 The electromagnetic energy–momentum tensor

Define $\tilde{T}_{\mathrm{E}i}{}^{j} := -F_{i}{}^{k}F_{k}{}^{j} + \frac{1}{4}F_{l}{}^{k}F_{k}{}^{l}\delta^{j}_{i}$ (or $\tilde{\mathbf{T}}_{\mathrm{E}} := -\mathbf{F}^2 + \frac{1}{4}(\mathrm{Tr}\mathbf{F}^2)\mathbf{I}$)

We shall calculate the quantity

$$\tilde{T}_{\mathrm{E}i}{}^{j}{}_{,j} = -F_{i}{}^{k}{}_{,j}F_{k}{}^{j} - F_{i}{}^{k}J_{k} + \tfrac{1}{2}F_{l}{}^{k}F_{k}{}^{l}{}_{,i} \tag{30}$$

From (23) we have

$$\begin{aligned} F_{l}{}^{k}F_{k}{}^{l}{}_{,i} &= -F_{l}{}^{k}(F^{l}{}_{i,k} + F_{ik,}{}^{l}) \\ &= F_{l}{}^{k}F_{i}{}^{l}{}_{,k} + F^{k}{}_{l}F_{ik,}{}^{l} \\ &= -2F_{k}{}^{j}F_{i}{}^{k}{}_{,j} \end{aligned}$$

This shows that the first and third terms on the right of (30) cancel, so that (30) becomes simply

$$\begin{aligned} \tilde{T}_{\mathrm{E}i}{}^{j}{}_{,j} &= -F_{i}{}^{k}J_{k} \\ &= -F_{i}{}^{k}4\pi e n'^{0} v_{k} \qquad \text{from (20)} \end{aligned}$$

where **v** is the velocity of the particles producing the field

$$= -4\pi n'^{0}\mathrm{d}P_{i}/\mathrm{d}s = -4\pi n^{0}\mathrm{d}P_{i}/\mathrm{d}x^{0}$$

from (14), using $m\mathrm{d}x^{i}/\mathrm{d}s = P^{i}$, (9) and (2).

Thus, apart from the factor -4π, the tensor divergence $\tilde{T}_{\mathrm{E}i}{}^{j}{}_{,j}$ gives the rate of change of momentum multiplied by the particle density.

Now let us consider a 'gas' of particles with charge in an electromagnetic field. Since the particles are no longer free we cannot conclude, as we did in §2.2.3, that the momentum in a spatial volume V can change only by particles leaving the volume; instead we now have

$$\begin{aligned} \frac{\mathrm{d}}{\mathrm{d}x^{0}}\int_{V} n^{0}P_{i}\mathrm{d}V &= -(\text{flux of } P_{i} \text{ out of } V) + \int_{V} n^{0}\mathrm{d}P_{i}/\mathrm{d}x^{0}\cdot\mathrm{d}V \\ &= -\text{flux} - \frac{1}{4\pi}\tilde{T}_{\mathrm{E}i}{}^{j}{}_{,j} \end{aligned}$$

If we repeat the argument of § 2.2.3 with this additional term included, then we find that (11) is now replaced by

$$T_{i}^{j}{}_{,j} = -\frac{1}{4\pi}\tilde{T}_{\mathrm{E}i}{}^{j}{}_{,j}$$

Thus if we define

$$T_{\text{total}} = T + \frac{1}{4\pi}\tilde{T}_{\mathrm{E}}$$

we shall then have

$$(T_{\text{total}})_{i}{}^{j}{}_{,j} = 0 \tag{31}$$

The energy–momentum tensor thus still satisfies the conservation law, provided that the electromagnetic field is included in the appropriate way. Thus we identify the relevant components of the tensor $T_E := (1/4\pi)\tilde{T}_E$ as the energy, momentum and fluxes associated with the electromagnetic field itself, and contributing additively to the total conserved energy–momentum. It is this total tensor that enters into the equations of general relativity in the next chapter.

Exercises

In questions 1 and 2 a light-wave is specified by a vector potential proportional to $\cos(k_i x^i)$, in some inertial frame **x**, where $\mathbf{k} = \omega(-1, \cos\theta, \sin\theta, 0)$. Here ω is the frequency of the light and θ is the angle made by the direction of propagation with the x^1-axis. k is, of course, a null covector.

1 Show that when one uses an inertial frame $\mathbf{x}'$ related to **x** by a boost with velocity v in the (x^0, x^1)-plane, the frequency and angle of the light change to ϱ' and θ', where

$$\omega' = \omega\beta(1 - v\cos\theta)$$
$$\cot\theta' = \cot\theta - v\,\mathrm{cosec}\,\theta$$

A distant star, lying in the plane of the Earth's orbit round the Sun and stationary in an inertial frame with the Sun as spatial origin, is emitting light with frequency ω. Neglecting terms in the square and higher powers of the Earth's orbital velocity v, investigate the change throughout the year of (a) the apparent position of the star and (b) the frequency of its light.

2 According to quantum theory, electromagnetic radiation manifests itself in discrete 'packets' called photons, which behave like particles of momentum $\hbar k$ (where k is the covector defined above and $2\pi\hbar$ is a fundamental quantity, Planck's constant).

A photon with frequency ω and with 3-momentum along the x^1-axis, in a frame **x**, strikes an electron of mass m, at rest at the spatial origin. The photon is reflected at a new frequency ω_2 and an angle θ to the x^1-axis, while the electron recoils at an angle ψ, say, to this axis. Show by considering the components of the conserved total momentum that

$$\frac{\omega_1 - \omega_2}{\omega_1\omega_2} = \frac{\hbar}{m}(1 - \cos\theta)$$

3 Two streams of particles flow in opposite directions, in some inertial frame **x**, one stream being made up of particles of mass m_1 and velocity v_1 in the positive x^1-direction, and the other having particles of mass m_2 and velocity $-v_2$. Show that an observer moving along the x^1-axis with velocity V will observe no net flux of momentum relative to himself provided that

$$\frac{V}{1+V^2} = \frac{v}{1+v^2}\cdot\frac{(\rho_1-\rho_2)}{(\rho_1+\rho_2)}$$

where ρ_1 and ρ_2 are the energy densities of the streams, i.e. $\rho_a = m_a\beta(v_a){n_a}^0$, n_a being the flux-vector of stream $a(a = 1, 2)$.

4 A particle of mass m and charge e moves in the circular orbit defined by $x^1 = R\cos\theta$, $x^2 = R\sin\theta$ (R a constant, $\theta \equiv \theta(s)$ where s is proper time), under the influence of cylindrically symmetric magnetic and electric fields **B** and **E**, **B** being in the x^3-direction and **E** being tangential to the orbit, both independent of θ. By considering $\mathrm{d}^2x^1/\mathrm{d}s^2$ and $\mathrm{d}^2x^2/\mathrm{d}s^2$ at $\theta = 0$, show that this is only possible if the magnitudes of the fields satisfy $E = -R\mathrm{d}B/\mathrm{d}t$.

5 Show that the equation of motion of a charged particle can be written

$$\dot{\mathbf{v}} + (v\dot{v}/1-v^2)\mathbf{v} = (e/m)(\mathbf{v}\wedge\mathbf{B}+\mathbf{E})$$

where a dot denotes derivative with respect to x^0 and v is the magnitude of the 3-vector **v**.

6 Choose a coordinate system **x** and define

$$x \ll y \quad \text{for} \quad x^0(x) < x^0(y) \quad \text{and} \quad \overrightarrow{xy} \quad \text{timelike}$$

$$x \prec y \quad \text{for} \quad x^0(x) \leq x^0(y) \quad \text{and} \quad \overrightarrow{xy} \quad \text{timelike or null.}$$

If $x \prec y$, define $\mathrm{d}(x, y) := (-g(\overrightarrow{xy}, \overrightarrow{xy}))^{\frac{1}{2}}$.
Prove the following:
(i) the relations $\prec$ and $\ll$ on M are unchanged if we use coordinates $\mathbf{x}' = \mathbf{Lx}$ with $\mathbf{L} \in \mathscr{L}^{\uparrow}$ (Exercise 8 of chapter 1);
(ii) $x \ll y$ and $y \prec z \Rightarrow x \ll z$
(you may find it easier to use coordinates with x and z on the time axis);
(iii) if $x \prec y \prec z$ then $[\mathrm{d}(x, y)+\mathrm{d}(y, z)] \leq \mathrm{d}(x, z)$;
(iv) if $x_0 \prec x_1 \prec x_2 \prec \ldots \prec x_n$ then $\sum_{r=1}^{n} \mathrm{d}(x_{r-1}, x_r) \leq \mathrm{d}(x_0, x_n)$
(i.e. a straight line is the longest curve between two points!)

3
General relativity

3.1 Freely falling coordinates

Perhaps the most famous discovery in physics, due to Galileo, is that all bodies fall with the same acceleration—or rather, that they would so fall if they were free from impedance by the air. This fact sets gravitation apart from all other forces and is responsible for the rather strange mathematical structure of general relativity.

We shall first consider a very delicate modern counterpart to Galileo's experiments on falling and rolling weights.

3.1.1 The Eötvös experiment

Imagine two weights on the ends of a light rod suspended in equilibrium from a fibre that can exert a restoring torque when it is twisted (a torsion balance) as shown in Figure 4. The weights are moving round with the rotation of the Earth, the inward acceleration that they need for this being caused by the combined effects of the gravitational forces, the tension T of the fibre and the torque or couple that it exerts, Γ.

To provide the acceleration needed for the circular motion, the weights require inward forces proportional to their masses, m_1 and m_2. Suppose, however, that the gravitational forces f_1 and f_2 are not quite

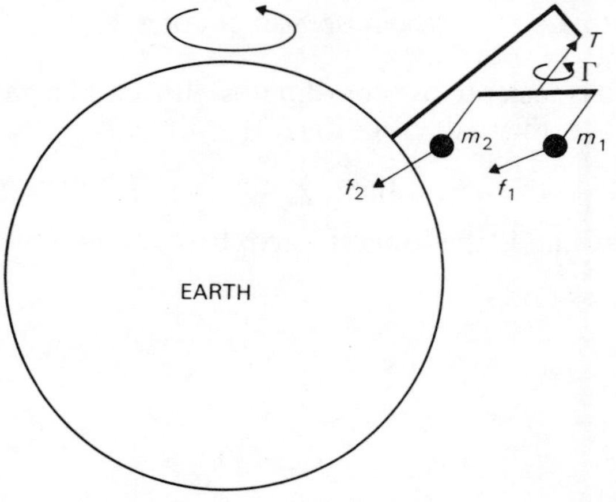

Figure 4. The Eötvös experiment.

in this proportion (if, for instance, the weights are made out of different substances). This would be equivalent to the weights having different accelerations if released. Then the difference between the gravitational forces and the total forces needed must be provided by a couple produced by a twist in the fibre (as far as the components perpendicular to the fibre are concerned).

To try to detect this twist, if it exists, suppose we rotate the clamp holding the fibre through exactly 180° so that the positions of the two masses are interchanged. If there were no twist due to out-of-balance gravitational forces, then the rod would also rotate through this angle. But if the gravitational forces are not proportional to the masses, then the torque that the fibre must supply to compensate will be reversed when the weights are interchanged (assuming that the rod is small so that the gravitational force on any one weight does not alter with its small change of position). So the fibre has to twist in the opposite direction in the new position, and the new equilibrium position of the rod will differ from 180° relative to the old by twice this twist of the fibre.

In the experiment as it was first performed by Baron von Eötvös around 1890, the whole apparatus—stand, fibre, weights, a housing to shield from draughts and a telescope to observe the position of the rod—was rotated through 180° and observations were made to see whether, at the new equilibrium, the rod was still in the same position relative to the apparatus. No significant difference was found, indicating that the gravitational forces on the weights were proportional to their masses, or that the weights would have the same accelerations if released.

In 1962–3 a modified experiment was performed by Dicke at Princeton. This time the apparatus remained fixed relative to the Earth, but use was made of the gravitational attraction of the Sun, whose direction relative to the laboratory moves round throughout the day. If the Sun's attraction affected the two weights differently there would, by the same argument, be an induced twist in the fibre which would vary with a 24-hour period. No periodic motion* was found. The best data, obtained when the experiment was repeated with improved accuracy by Braginsky and Panov in 1970, suggest that gravity imparts the same acceleration to all bodies, to an accuracy of *one part in* 10^{12}.

3.1.2 The equivalence principle

Imagine a laboratory contained in an artificial satellite in free fall, acted on by gravity only. The laboratory and everything in it will then have

*Actually the rod, in this case a quartz frame, was held stationary relative to the apparatus by applying an electrical force; the force needed to do this was then examined for any 24-hour periodicity.

precisely the same acceleration, neglecting for the moment the variation in gravity from one part of space to another. So there will be no relative acceleration at all: relative to the laboratory everything will be 'weightless'. The Eötvös experiment puts an upper limit on the apparent 'weight' of a body in the laboratory of $1/10^{12}$ of the terrestrial weight. This is so accurately zero that we believe it to be a basic physical principle that gravity is precisely annulled by the acceleration of the satellite towards the earth. Although the Eötvös experiment only shows directly that the dynamical effect of gravity would be annulled, it is reasonable to postulate the principle for all effects of gravity. If there were some effect of gravity (such as an electromagnetic effect) that was not annulled by free fall, then this effect would almost inevitably alter the body's energy, and hence by special relativity its mass, which would result in a dynamical effect that would be detectable by the Eötvös experiment.

This means that a freely falling observer can construct a coordinate system, which we shall call *freely falling coordinates*, in which he is at rest at the spatial origin and in which, near the spatial origin, the laws of special relativity are valid with no gravitational forces needed. This statement is one form of the equivalence principle (or 'strong' equivalence principle).

The name comes from the converse proposition: since the frame fixed on Earth where gravitational forces are experienced is accelerating relative to a freely falling frame of reference, we see that gravitational effects are *equivalent* to the effects of an acceleration relative to a freely falling frame. Gravity is simply an expression of a state of acceleration relative to freely falling coordinates.

Of course, the concept of freely falling coordinates is rather vague. In the course of developing our mathematical model it will be translated into the more precise ideas of a freely falling frame (§ 3.3.1) and Fermi–Walker coordinates (§ 3.3.12).

3.1.3 General coordinates

In special relativity we used inertial frames related by Poincaré transformations. General relativity replaces these by freely falling coordinates; what, then, are the corresponding transformations? Here lies the essentially new feature of general relativity.

Because gravity varies from place to place, there can be no coordinate system in which special relativity, with no gravitational forces, is valid everywhere. If there were such a system, there would be no such thing as gravitation. The best that can be achieved is a coordinate system in which special relativity is valid in the limit as the world-line of some particular observer is approached. In a sense that will later be made precise, the coordinates are only determined close to the observer, and further away they can be chosen arbitrarily, no one choice being better than another. Instead of a neat group of inertial frames we must allow for arbitrary coordinates away from the world-line, and so we are involved in quite general coordinate transformations from one general set of coordinates to another, without the linearity seen in special relativity.

The coordinate transformation expressing $x^0, \ldots, x^3$ as functions of $x'^0, \ldots,$ are now merely four arbitrary functions, mapping one region of $\mathbb{R}^4$ (the range of the $\mathbf{x}'$-coordinates) into another (the range of the $\mathbf{x}$-coordinates). Of course, different points must always be labelled by different coordinates so that the transformation must be one–one. We shall make the further mathematical assumption that the functions involved can be differentiated as often as necessary, and we can arrange that the matrix with components $\partial x^i/\partial x'^j$ is non-singular. The need for this will emerge shortly.

To summarize the new picture: space–time M, the set of all possible 'events', is now simply a topological space in which every event has a neighbourhood that can be given coordinates in a one–one fashion; the map taking x to its coordinates $\mathbf{x}(x)$ is continuous, and so is its inverse, and different coordinates are related differentiably as described. This is essentially the definition of M as a *differentiable manifold*.

3.2 Tangent vectors and tensors

We must now return to the equivalence principle in a mathematical form by defining the concept of 'close to the observer' used in §§ 3.1.2, 3. Let x be an event on an observer's world-line. We shall construct a vector space $T_x(M)$ that corresponds to 'infinitesimal' displacements from x. The whole of special relativity, described in terms of $T.(M)$ in chapter 1, is then translated completely into $T_x(M)$, expressing the idea that special relativity is valid 'infinitesimally near' x. The definitions used will closely parallel those of § 1.4.1.

3.2.1 The tangent vector to a curve

A *differentiable curve* is a continuous map γ from an interval (a, b) of the real line into M, such that the functions $x^i(\gamma(u))$ are differentiable in u for some (and hence for any) coordinates.

Two differentiable curves γ_1 and γ_2 are *tangent* at u_0 if

(i) $\gamma_1(u_0) = \gamma_2(u_0)$
(ii) for some (and hence any) coordinates

$$\left.\frac{\mathrm{d}\mathbf{x}(\gamma_1(u))}{\mathrm{d}u}\right|_{u_0} = \left.\frac{\mathrm{d}\mathbf{x}(\gamma_2(u))}{\mathrm{d}u}\right|_{u_0}$$

The *tangent vector* to γ at the event $x = \gamma(u_0)$ is the equivalence class of all curves tangential to γ at u_0, and is written $\dot{\gamma}(u_0)$. Clearly, given a coordinate system **x**, a tangent vector is uniquely specified by the set of components

$$\mathbf{X} = \begin{pmatrix} X^0 \\ X^1 \\ X^2 \\ X^3 \end{pmatrix}$$

where $X^i = (\mathrm{d}/\mathrm{d}u)x^i(\gamma(u_0))$ for some γ defining the tangent vector.

To justify thinking of tangent vectors as infinitesimal displacements we need only recall the definition of a derivative

$$\mathbf{X} = \lim_{h \to 0} \frac{1}{h}[\mathbf{x}(\gamma(u_0 + h)) - \mathbf{x}(\gamma(u_0))]$$

Thus **X** is constructed from a displacement along the curve from $\gamma(u_0)$ to $\gamma(u_0 + h)$, with the limit taken as this displacement shrinks to zero—precisely what is needed as the analogue of the finite displacements that define vectors in special relativity.

3.2.2 Coordinate transformations

We now compare the coordinates **X** and **X**$'$ of a given tangent vector $X = \dot{\gamma}(u_0)$ in coordinates **x** and **x**$'$ respectively. Putting $x_0 := \gamma(u_0)$, then, by the chain rule for derivatives,

$$X'^i = \frac{\mathrm{d}}{\mathrm{d}u} x'^i(\gamma(u))\bigg|_{u_0} = \frac{\partial x'^i}{\partial x^j}\bigg|_{x_0} \frac{\mathrm{d}}{\mathrm{d}u} x^j(\gamma(u))\bigg|_{u_0}$$

or

$$X'^i = L^i{}_j X^j \tag{1}$$

where

$$L^i{}_j := \partial x'^i / \partial x^j |_{x_0} \tag{2}$$

(Note that this is identical to the transformation given by (9) of chapter 1 for special relativity, except that now L could be any non-singular matrix. Because of this non-singular condition, which we imposed in §3.1.3, the correspondence between **X** and **X**$'$ is one–one, and hence the definition of tangent vectors given above is independent of the choice of coordinates.)

3.2.3 The tangent space and tensors

Every differentiable curve through an event x defines a tangent vector to the curve at x. We write $T_x(M)$ for the set of all such tangent vectors at x.

$T_x(M)$ is vector space: addition $(X+Y)$ and multiplication by a scalar (aX) are defined by performing corresponding operations on the components, $\mathbf{X}+\mathbf{Y}$ and $a\mathbf{X}$. Since (1) is linear, this definition of the operations is independent of the coordinates, just as was the case for $T.(M)$ in §1.4.

Associated with $T_x(M)$ is its dual $T_x^*(M)$: the space of real linear functions on $T_x(M)$. Thus if ω belongs to $T_x^*(M)$, then for any coordinates $\mathbf{x}$ we can find a row vector $\boldsymbol{\omega}$ of components for ω so that

$$\omega(X) = \boldsymbol{\omega}\mathbf{X}$$

for all X in $T_x(M)$, just as in §1.4.2. The definition of the basis vectors $\underset{0}{E}, \ldots, \underset{3}{E}$ used in this connection remains unchanged.

Moreover, equation (12) of chapter 1 remains valid: under a change of coordinates

$$\boldsymbol{\omega}' = \boldsymbol{\omega}\mathbf{L}^{-1} \tag{3}$$

A useful version of this can be derived from the following.

Lemma $(L^{-1})^i{}_l = \partial x^i/\partial x'^l$

Proof Clearly $\delta^i_j = \partial x^i/\partial x^j \qquad (= 0 \text{ unless } i = j, \text{ when } = 1)$

$$= \frac{\partial x^i}{\partial x'^k}\frac{\partial x'^k}{\partial x^j} \qquad \text{(chain rule)}$$

$$= \frac{\partial x^i}{\partial x'^k} L^k{}_j \qquad \text{(from (2))}$$

Multiplying both sides by $(L^{-1})^j{}_l$ and summing over j now gives the result.

This enables us to write (3) as

$$\omega'_l = \omega_i \frac{\partial x^i}{\partial x'^l} \tag{4}$$

Just as in §1.6, we can form tensor products of vectors and covectors to form tensors: r copies of $T_x(M)$ and s copies of $T_x^*(M)$ give, under tensor multiplication and linear operations, the vector space $T_x^{(r,s)}(M)$. The general tensor transformation law is then

$$T'_{ij\cdots l}{}^{pq\cdots t} = (L^{-1})^m{}_i (L^{-1})^n{}_j \ldots (L^{-1})^o{}_l L^f{}_u L^q{}_v \ldots L^t{}_w T_{mn\cdots o}{}^{uv\cdots w}$$

(cf. §1.7.2).

3.3 Mathematical modelling of the equivalence principle

3.3.1 Freely falling frames

Any set of four vectors $(\underset{0}{E}, \ldots, \underset{3}{E})$ that form a basis for $T_x(M)$ (i.e. that are linearly independent) is called a *frame* at x. In particular, a system of freely falling coordinates $\bar{\mathbf{x}}$ for an observer with world-line γ has a coordinate basis of vectors $\underset{i}{\bar{E}}$ defined by giving their components in the $\bar{\mathbf{x}}$-coordinates as $\underset{i}{\bar{E}}^j = \delta_i^j$ (see §1.4.1.). The family of frames defined in this way at all the points of γ by freely falling coordinates is called a freely falling frame on γ. Note that any vector X at a point of γ can be written $X = \bar{X}^i \underset{i}{\bar{E}}$ (cf. chapter 1, equation (10)), where $\bar{X}^i$ are the components of X in freely falling coordinates, and $(\underset{0}{\bar{E}}, \ldots, \underset{3}{\bar{E}})$ is the corresponding freely falling frame.

The fundamental principle, that special relativity holds in freely falling coordinates, means that there is a precise correspondence between the components of a general relativistic tensor in a freely falling frame and the components of the corresponding tensor in special relativity. For example, the energy–momentum tensor of special relativity (§2.2.2) corresponds to a tensor field T in general relativity, such that on the world-line of a freely falling coordinate system, the components $\bar{T}_{ij}$ are the mass-density, momentum density and stress components (§2.2.4).

More importantly, in special relativity we have the metric tensor, of type (0, 2). This is taken over into general relativity by supposing that at each point x_0 of M there is specified a tensor $g(x_0)$ in $T_{x_0}^{(0,2)}(M)$, and, in accordance with the equivalence principle, there exist coordinates $\bar{\mathbf{x}}$ near x_0 (namely, freely falling coordinates based on a freely falling observer whose world-line passes through x_0) such that the components of g at the point x_0 (but not necessarily elsewhere) in these coordinates take the special relativity form

$$\bar{\mathbf{g}}(x_0) = \begin{pmatrix} -1 & 0 & 0 & 0 \\ 0 & 1 & 0 & 0 \\ 0 & 0 & 1 & 0 \\ 0 & 0 & 0 & 1 \end{pmatrix} = \overset{0}{\mathbf{g}} \qquad \text{say}$$

Thus from now on g will denote a general metric (tensor field) on M with this property, and the special relativity metric will be distinguished as $\overset{0}{g}$. The space–time of special relativity, in which there are coordinates for which $\mathbf{g} = \overset{0}{\mathbf{g}}$ everywhere, will be called *Minkowski space.*

Transformation to freely falling coordinates at a particular point actually involves only the freely falling frame at that point. If $\bar{\mathbf{x}}$ denotes the freely falling coordinates, and T is a tensor of type $(0, s)$, then the $\bar{\mathbf{x}}$-coordinates of T are given by

$$\bar{T}_{ij\ldots} = \bar{T}_{pq\ldots}\,\delta^p_i\,\delta^q_j = T(\underset{i}{\bar{E}}, \underset{j}{\bar{E}}, \ldots)$$

And, in particular, the frame is related to the metric by

$$\overset{0}{g}_{ij} = g(\underset{i}{\bar{E}}, \underset{j}{\bar{E}})$$

More generally, given a tensor of type (r, s) with components $\bar{T}_{ij\ldots}{}^{lm\cdots}$ in the freely falling coordinates, we have

$$\bar{T}_{ij\ldots}{}^{lm\cdots} = \bar{T}_{pq\ldots}{}^{uv\cdots}\,\delta^p_i\,\delta^q_j\ldots\delta^l_u\,\delta^m_v\ldots$$

We now regard δ^l_u as the u-component of the covector $\overset{0}{g}{}^{lm}g(\underset{m}{\bar{E}})$, since $(\overset{0}{g}{}^{lm}g(\underset{m}{\bar{E}}))_u = \overset{0}{g}{}^{lm}(\bar{g}_{uv}\underset{m}{\bar{E}}{}^v) = \overset{0}{g}{}^{lm}\overset{0}{g}_{uv}\delta^v_m = \delta^l_u$, so that the above equation becomes

$$\bar{T}_{ij\ldots}{}^{lm\cdots} = \overset{0}{g}{}^{lp}\overset{0}{g}{}^{mq}\ldots T(\underset{i}{\bar{E}}, \underset{j}{\bar{E}}, \ldots, g(\underset{p}{\bar{E}}), g(\underset{q}{\bar{E}}), \ldots)$$

3.3.2 Freedom of choice of freely falling frames

At any one point the relation $g(\underset{i}{\bar{E}}, \underset{j}{\bar{E}}) = \overset{0}{g}_{ij}$ restricts the freedom of transformation to a different frame to those transformations for which the metric at the point still has the components $\overset{0}{g}_{ij}$, and this is, by definition, simply the Lorentz group. In other words, we could equally well use the frame $\underset{j}{\tilde{E}} := L^i_j\underset{i}{\bar{E}}$, provided that L is a member of the Lorentz group.

Now suppose we proceed to a freely falling frame defined all along the world-line of some observer. At any point on the line the freedom of tranformation is now restricted to those Lorentz transformations that keep $\underset{0}{\bar{E}}$ unchanged as the tangent vector to the world-line, and these are just spatial rotations (and reflections, see §1.3.3). Further, the choice of the frame at any one point in fact determines it all along the world-line.

To see this, imagine an observer in free fall in a spacecraft. He will be able to tell whether his craft is rotating relative to a freely falling frame by the presence or absence of centrifugal force. Thus if the spacecraft is to define correctly a freely falling frame, then its orientation, once chosen at some initial instant, is fixed for all subsequent time by requiring that there be no detectable rotation. Thus the only available freedom in the choice of a freely falling frame on the world-line of the observer is a transformation by a *constant* three-dimensional rotation matrix, corresponding to a different choice of the initial orientation of the spatial coordinates. (The time axis vector $\underset{0}{\bar{E}}$ is, of course, always proportional to the tangent vector to the world-line, scaled so that $g(\underset{0}{\bar{E}}, \underset{0}{\bar{E}}) = \overset{0}{g}_{00} = -1$, and so is fixed.)

While a freely falling *frame* is thus quite restricted, the freely falling *coordinates* can vary considerably as we move away from γ while still defining the same frame. If two coordinates $\bar{\mathbf{x}}$ and $\tilde{\mathbf{x}}$ satisfy

$$\frac{\partial \tilde{x}^i}{\partial \bar{x}^j} = \delta^i_j$$

on γ, then by (1) and (2) we see that the same $\underset{i}{\bar{E}}$ will constitute a freely falling frame on γ for both coordinates. Thus the above physical arguments only fix the derivatives of a freely falling coordinate system on γ, and when constructing such a coordinate system, as we shall do in §3.3.12, we must supplement the physical restriction on the freely falling frame by some mathematical condition that will determine how we are to continue the coordinates away from γ.

3.3.3 Rates of change

In special relativity the rate of change along a world-line was given by the directional derivative $\underset{X}{\nabla}$ in the direction of the tangent vector X (§1.9.3). The same holds true for the rate of change of a *function* in general relativity, for which we adopt the same notation and definitions as in special relativity. But we run into difficulty on trying to differentiate vector and tensor fields. Suppose, for instance, that we were to define the directional derivative $\underset{X}{\nabla} Y$, for a vector field Y and a vector X, by the components $X^j \partial Y^i / \partial x^j$ (as in special relativity). In a different coordinate system the corresponding expression would be

$$\begin{aligned} X'^j \frac{\partial Y'^i}{\partial x'^j} &= \left(\frac{\partial x'^j}{\partial x^k} X^k\right) \frac{\partial x^m}{\partial x'^j} \frac{\partial}{\partial x^m} \left(\frac{\partial x'^i}{\partial x^l} Y^l\right) \\ &= \delta^m_k X^k \left(\frac{\partial x'^i}{\partial x^l} \frac{\partial Y^l}{\partial x^m} + \frac{\partial^2 x'^i}{\partial x^l \partial x^m} Y^l\right) \\ &= \frac{\partial x'^i}{\partial x^l} \left(X^m \frac{\partial Y^l}{\partial x^m}\right) + X^m Y^l \frac{\partial^2 x'^i}{\partial x^l \partial x^m} \end{aligned} \tag{5}$$

Now, if this expression denoted the components of the same vector as defined in $\mathbf{x}$-coordinates, only the first term (a transformation by the correct matrix (2)) should be present. The presence of the second term shows that the suggested expression does not define a vector or, more accurately, defines a different vector in each coordinate system—and so cannot be used in constructing geometrically correct laws based on the vector space $T_x(M)$.

To find the correct expression we restrict attention to the rate of change along the world-line of a freely falling observer, parametrized by u with tangent vector X and freely falling frame $(\underset{i}{\bar{E}})$. If $Y = \bar{Y}^i \underset{i}{\bar{E}}$ is a vector field defined on and near γ, let us define its rate of change as

$$\frac{\mathrm{D}Y}{\mathrm{D}u} \equiv \underset{X}{\nabla} Y := \underset{i}{\bar{E}} \frac{\mathrm{d}\bar{Y}^i(\gamma(u))}{\mathrm{d}u} \tag{6}$$

valid *only* for a freely falling frame. This does not depend on the choice of such a frame, because, as we have just seen, the only available freedom is a fixed spatial rotation giving the transformation $\underset{i}{\bar{E}} \to R_i{}^j \underset{j}{\bar{E}}, \quad \bar{Y}^i \to (R^{-1})_i{}^j \bar{Y}^i$, under which the equation (6) is invariant.

We now assume* that the quantity $\underset{X}{\nabla} Y$, as thus defined, is linear in X; in other words, if through some event there run the world-lines of three observers with tangent vectors X, Z and $aX + bZ$, then we have

$$\underset{aX+bZ}{\nabla} Y = a\underset{X}{\nabla} Y + b\underset{Z}{\nabla} Y \tag{7}$$

From the definition (6) we deduce the properties of linearity in Y

$$\underset{X}{\nabla}(aY + bW) = a\underset{X}{\nabla} Y + b\underset{X}{\nabla} W \tag{8}$$

and also *Leibnitz' rule*, that, for a real function f

$$\underset{X}{\nabla}(fY) = (\underset{X}{\nabla} f)Y + f\underset{X}{\nabla} Y \tag{9}$$

The vector $\underset{X}{\nabla} Y$ is called the *covariant derivative* of Y in the direction X. Our argument has shown how it is needed in order to express the idea of a freely falling observer measuring rates of change. We shall now construct the mathematical model to be used for general relativity by taking as our starting point the space–time M described at the end of §3.1.3, the tangent spaces and tensor spaces developed in §3.2, and an operation of covariant differentiation satisfying (7), (8) and (9) above. Arguing in the reverse direction to that taken so far in this section, we shall show that these basic structures can be used to fix uniquely the curves that we regard as world-lines of freely falling observers, and the

* This can be deduced from the requirement that when three world-lines form a small triangle, the successive changes in Y along two of the sides add up to the change in Y along the third.

freely falling frames on them. Before doing this, we first show how to express the covariant derivative in terms of certain coordinate-dependent quantities.

3.3.4 The connection

Now let us go over to an arbitrary (not freely falling) coordinate system, and let $\underset{i}{E}$ be the coordinate basis vectors.

If Y is a vector field having components Y^i then

$$\underset{X}{\nabla} Y = \underset{X}{\nabla}(Y^k \underset{k}{E}) = (\underset{X}{\nabla} Y^k)\underset{k}{E} + Y^k(\underset{X}{\nabla}\underset{k}{E}) \qquad \text{(from (8), (9))}$$

From the definition of the directional derivative of a function given in chapter 1, equation (23), we have

$$\underset{X}{\nabla} Y^k = X^j Y^k{}_{,j}$$

Also

$$\underset{X}{\nabla}\underset{k}{E} = \underset{(X^j \underset{j}{E})}{\nabla}\underset{k}{E} = X^j \underset{\underset{j}{E}}{\nabla}\underset{k}{E} \qquad \text{(from (7))}$$

$$= X^j \Gamma^i_{kj} \underset{i}{E}$$

where

$$\Gamma^i_{kj} := (\underset{\underset{j}{E}}{\nabla}\underset{k}{E})^i \tag{10}$$

Hence

$$\underset{X}{\nabla} Y = (X^j Y^k{}_{,j})\underset{k}{E} + Y^k(X^j \Gamma^i_{kj} \underset{i}{E})$$

$$= X^j (Y^i{}_{,j} + \Gamma^i_{kj} Y^k)\underset{i}{E} \tag{11}$$

(relabelling indices).

The set of quantities Γ^i_{kj}, which determines the operation of covariant differentiation, is called the *connection* (strictly the linear, or in older texts affine, connection). It is not a tensor: on transformation, a term involving second derivatives of the coordinates is added, which compensates for the extra term in (5) so that the whole expression (11) is a vector. The full transformation law can be deduced from (5) (see Exercise 4).

3.3.5 Freely falling frames and geodesics

The covariant derivative is linked to the idea of free fall by (6). If Y in this equation is taken to be one of the coordinate basis vectors $\underset{i}{\bar{E}}$ for some freely falling coordinates $\bar{\mathbf{x}}$ (we continue to use a plain $\mathbf{x}$ for a general coordinate system), then, since the components of $\underset{i}{\bar{E}}$ in these coordinates are the constant numbers δ_i^j we have

$$\frac{\mathrm{D}\underset{i}{\bar{E}}}{\mathrm{D}u} \equiv \nabla_{\dot{\gamma}} \underset{i}{\bar{E}} = 0 \tag{12}$$

Now, applying the ideas of §1.8.1, we can always alter the parameter on γ to a new parameter s, called proper time, so that the new tangent vector is $\underset{0}{\bar{E}}$ (instead of merely being proportional to it). So we now apply (12) to $\underset{0}{\bar{E}} = \dot{\gamma}$.

Using (11) to evaluate (12) in this case we get

$$0 = \nabla_{\dot{\gamma}} \underset{0}{\bar{E}} = \nabla_{\dot{\gamma}} \dot{\gamma} = \dot{\gamma}^j(\dot{\gamma}^i{}_{,j} + \Gamma^i_{kj}\dot{\gamma}^k)\underset{i}{E} \tag{13}$$

Now $\dot{\gamma}^i = \mathrm{d}x^i/\mathrm{d}s$, where s is the proper time parameter on the world-line, and the first term of (13) is

$$\dot{\gamma}^j\dot{\gamma}^i{}_{,j} = \nabla_{\dot{\gamma}}\left(\frac{\mathrm{d}x^i}{\mathrm{d}s}\right) = \frac{\mathrm{d}^2x^i}{\mathrm{d}s^2}$$

thus (13) becomes

$$\frac{\mathrm{d}^2x^i}{\mathrm{d}s^2} + \Gamma^i_{kj}\frac{\mathrm{d}x^k}{\mathrm{d}s}\frac{\mathrm{d}x^j}{\mathrm{d}s} = 0 \tag{14}$$

This gives an equation specifying the world-line of a freely falling observer in terms of the connection. Any solution of the equation is called a *geodesic*.

Thus to determine what lines are candidates for freely falling observers, we simply solve this second-order differential equation. Not all geodesics are possible world-lines: later, the metric will be used to classify geodesics into timelike, null and spacelike types, of which only the first can be the world-line of an observer. For any given geodesic, a freely falling frame is then given by the equation (12) that determines the frame for all s, once its initial orientation is specified.

We can use the notation of (12) to write the geodesic equation in another form

$$\frac{\mathrm{D}\dot{\gamma}}{\mathrm{D}s} = 0$$

This defines a geodesic as a curve whose tangent vector is *constant* (zero covariant derivative) as we proceed along it; expressed loosely, it always goes in the same direction. In this sense, a geodesic is the generalization of the concept of a *straight line*. Physically this is quite understandable: a freely falling observer feels as though he is proceeding with uniform velocity in a straight line, or is in a state of rest.

However, the geometry of space–time on a larger scale makes geodesics behave in a very different way from the straight lines of Euclidean geometry.

An important feature of (14) is that it allows a certain amount of reparametrization of geodesics. Suppose we define $\tilde{\gamma}$ by

$$\tilde{\gamma}(u) = \gamma(au+b) \tag{15}$$

corresponding to the substitution

$$s = au+b \tag{16}$$

where a and b are constants. Writing $\mathbf{x}(u)$ for $\mathbf{x}(\tilde{\gamma}(u))$ we have

$$\begin{aligned}\Gamma^i_{kj}\frac{\mathrm{d}x^k}{\mathrm{d}u}\frac{\mathrm{d}x^j}{\mathrm{d}u}+\frac{\mathrm{d}^2x^i}{\mathrm{d}u^2} &= \Gamma^i_{kj}\left(\frac{\mathrm{d}s}{\mathrm{d}u}\right)^2\frac{\mathrm{d}x^k}{\mathrm{d}s}\frac{\mathrm{d}x^j}{\mathrm{d}s}+\frac{\mathrm{d}s}{\mathrm{d}u}\frac{\mathrm{d}}{\mathrm{d}s}\left(\frac{\mathrm{d}x^i}{\mathrm{d}s}\frac{\mathrm{d}s}{\mathrm{d}u}\right)\\ &= a^2\left(\Gamma^i_{kj}\frac{\mathrm{d}x^k}{\mathrm{d}s}\frac{\mathrm{d}x^j}{\mathrm{d}s}+\frac{\mathrm{d}^2x^i}{\mathrm{d}s^2}\right) = 0\end{aligned}$$

So that $\tilde{\gamma}$ is again a geodesic. It is easy to develop this argument to show that (16) is in fact the most general reparametrization for which $\tilde{\gamma}$ still satisfies the geodesic equation. Any new parameter that has this property is called an *affine parameter.*

3.3.6 Parallel transport

Now let γ be any curve (freely falling or not) through an event $x = \gamma(0)$, and let $Y \in T_x(M)$ be a vector at x. Then we can define a vector field $Y(u)$ along γ (although strictly speaking it should not be called a field, being defined only on a curve) by requiring

$$\left(\frac{\mathrm{D}Y}{\mathrm{D}u}\right)^i = \frac{\mathrm{d}Y^i}{\mathrm{d}u}+\Gamma^i_{kj}\,Y^k\dot{\gamma}^j = 0$$

$$Y(0) = Y$$

This is a differential equation with an initial condition that determines $Y(u)$ uniquely (providing that Γ^i_{kj} are sufficiently smooth). We think of the resulting solution as arising from 'carrying' the vector Y along the curve, keeping it 'constant' throughout, in the sense of having zero covariant derivative. Such a vector field is said to be parallely transported (or propagated).

In this way, if we are given two events x and y, a curve γ joining the two and a vector Y at x, then we can transport Y along γ to y by forming $Y(u)$ and noting its value at the event y. However, it is vital that we do not think of the parallely propagated vector at y as being 'the same' as the vector at x, because in general if we were to use a different curve along which to transport the vector to y, then we would produce a different vector at y: parallel transport from one point to another depends on the route taken.

3.3.7 A two-dimensional analogue

To fix our ideas and make the concepts easier to visualize, let us consider a curved two-dimensional surface instead of space–time. All our formalism can be applied except that indices will now run over the values $(1, 2)$. Suppose our surface is the unit sphere in $\mathbb{R}^3$ with polar coordinates $x^1 = \theta$, $x^2 = \phi$.

We can imagine the coordinate basis vectors $\underset{1}{E}$ and $\underset{2}{E}$ as arising from small displacements $\delta\theta$ and $\delta\phi$ along the curves $\phi =$ constant and $\theta =$ constant respectively (see Figure 5).

We shall now produce and discuss a connection that corresponds to our intuitive notion of 'parallel transport' (it can be derived by methods developed later: see Exercise 3). Let us demand that

$$\nabla_{\underset{1}{E}} \underset{1}{E} = 0 \tag{17a}$$

$$\nabla_{\underset{1}{E}} \underset{2}{E} = \cot\theta \underset{2}{E} = \Gamma^2_{21} \underset{2}{E} \tag{17b}$$

$$\nabla_{\underset{2}{E}} \underset{1}{E} = \cot\theta \underset{2}{E} = \Gamma^2_{12} \underset{2}{E} \tag{18a}$$

$$\nabla_{\underset{2}{E}} \underset{2}{E} = -\sin\theta\cos\theta \underset{1}{E} = \Gamma^1_{22} \underset{1}{E} \tag{18b}$$

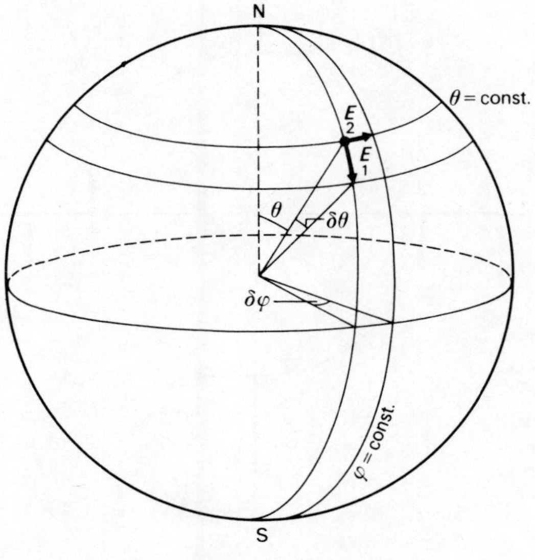

Figure 5

To justify this, consider first the pair (17). These refer to variation along the curves ϕ = constant (tangent vector $\underset{1}{E}$). Then (17a) can be interpreted as meaning that the vector $\underset{1}{E}$ stays 'the same' as we move along these curves, which is reasonable.

If we note that $\underset{\underset{1}{E}}{\nabla} \sin\theta = (\mathrm{d}/\mathrm{d}\theta)\sin\theta = \cos\theta$, then (17b) can be written as

$$0 = \underset{\underset{1}{E}}{\nabla}\underset{2}{E} - \left(\frac{1}{\sin\theta}\underset{\underset{1}{E}}{\nabla}\sin\theta\right)\underset{2}{E} = \sin\theta\,\underset{\underset{1}{E}}{\nabla}\left(\frac{1}{\sin\theta}\underset{2}{E}\right)$$

meaning that the vector $(1/\sin\theta)\underset{2}{E}$ is 'constant' along the curves ϕ = constant. Again, this is reasonable because the vector $\underset{2}{E}$ shrinks to zero at the poles, and the factor $(1/\sin\theta)$ compensates for this.

The pair (18) are rather more opaque, but they become clear in the two extreme cases $\theta = \pi/2$ and $\theta = \varepsilon \approx 0$. On $\theta = \pi/2$, the equator, the right-hand-sides of (18a) and (18b) vanish, expressing the 'constancy' of both $\underset{1}{E}$ and $\underset{2}{E}$ round the equator.

On the other hand, if we traverse a small circle near the North Pole,

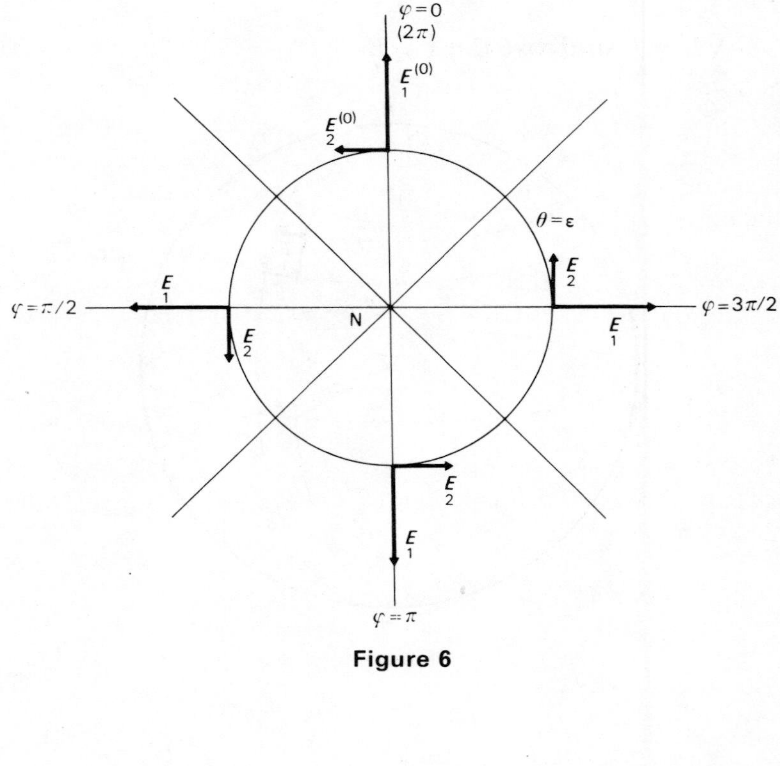

Figure 6

the vector $\underset{1}{E}$ and the rescaled vector $(1/\sin\theta)\underset{2}{E}$ rotate as we go round (Figure 6). If we were in ordinary flat space, and not on a curved surface, we could describe this rotation by

$$\underset{1}{E} = \underset{1}{E}^{(0)}\cos\phi + (1/\sin\theta)\underset{2}{E}^{(0)}\sin\phi \tag{19}$$

$$(1/\sin\theta)\underset{2}{E} = (1/\sin\theta)\underset{2}{E}^{(0)}\cos\phi - \underset{1}{E}^{(0)}\sin\phi$$

where $\underset{1}{E}^{(0)}$ and $\underset{2}{E}^{(0)}$ are the positions of the vectors at $\phi = 0$. Actually this is not valid, because the vectors for different values of ϕ are in different tangent spaces and so cannot be equated in this way; however, a very small region near the pole should behave like flat space, and we should expect similar equations to hold. Differentiating (19) with respect to ϕ gives

$$\nabla_{\underset{2}{E}}\underset{1}{E} = \frac{D\underset{1}{E}}{D\phi} = -\underset{1}{E}^{(0)}\sin\phi + (1/\sin\theta)\underset{2}{E}^{(0)}\cos\phi$$

$$= (1/\sin\theta)\underset{2}{E} \tag{20}$$

$$\nabla_{\underset{2}{E}}\underset{2}{E} = \frac{D\underset{2}{E}}{D\phi} = \sin\theta[-(1/\sin\theta)\underset{2}{E}^{(0)}\sin\phi - \underset{1}{E}^{(0)}\cos\phi]$$

$$= -\sin\theta\,\underset{1}{E} \tag{21}$$

which is indeed the limiting form of (18) as $\phi \to 0$.

The connection just described seems the most 'natural' one to place on a sphere (and we shall see later that it is mathematically distinguished from other connections), but we are not forced to use it. For instance, as discussed by Schouten in 'Ricci Calculus', if we wished to describe the motion on the Earth of a man who always faced a fixed point then we could take coordinates with one of the poles at this point and require that the vector $\underset{1}{E}$, pointing to the pole, was always parallely transported. For this the connection could be taken to be given by $\nabla_X \underset{1}{E} = \nabla_X (1/\sin\theta)\underset{2}{E} = 0$ for every X. (This would be one case where parallel transport did not depend on the path taken.)

In the following sections we shall show how to select one particular connection on space–time as the 'natural' one, by using the metric and the idea of torsion.

3.3.8 The metric

We now return to the metric tensor, introduced in §3.3.1.

As usual, we assume for mathematical convenience that the components $g^{ij}(\mathbf{x})$ are sufficiently smooth functions of the coordinates.

We shall continue to use the convention of raising and lowering indices (§1.6.1), now using the general metric g for this purpose. As before, the inverse of g has components written g^{ij}. The only difference concerns the transformation matrix $L^i{}_j = \partial x'^i/\partial x^j$. It can be seen that the index i belongs to the x'-coordinates, while j belongs to the x-coordinates. (Some authors indicate this by writing $L^{i'}{}_j$.) Thus one index should be raised with g and the other lowered with g'. As this is liable to cause confusion we shall not apply the convention to $L^i{}_j$, thus we shall always write $(L^{-1})^i{}_j$ (not $L_i{}^j$) or the equivalent explicit expression $\partial x^i/\partial x'^j$.

3.3.9 Commutators and torsion

Partial derivatives can be taken in any order, so that $f_{,ij} = \partial^2 f/\partial x^i \partial x^j = f_{,ji}$. However, it is not true, even in special relativity, that $\underset{X}{\nabla}\underset{Y}{\nabla} f = \underset{Y}{\nabla}\underset{X}{\nabla} f$. From the definition equation (23) of chapter 1, retained in general relativity for differentiating functions, we have

$$\underset{X}{\nabla}\underset{Y}{\nabla} f = X^i(Y^j f_{,j})_{,i} = X^i Y^j{}_{,i} f_{,j} + X^i Y^j f_{,ji}$$

and so

$$\underset{X}{\nabla}\underset{Y}{\nabla} f - \underset{Y}{\nabla}\underset{X}{\nabla} f = \underset{[X,Y]}{\nabla} f \tag{22}$$

where

$$[X,Y]^j := X^i Y^j{}_{,i} - Y^i X^j{}_{,i} \tag{23}$$

is called the *commutator* of X and Y. It is a vector (even though it involves partial derivatives of vectors) because the right-hand-side of (22) is the function $\nabla f([X,Y])$, where ∇f is a covector, (cf. the proposition of §1.7).

A closely related vector is the quantity $\underset{X}{\nabla} Y - \underset{Y}{\nabla} X$; we shall denote the difference between this and $[X,Y]$ by

$$S(X,Y) := \underset{X}{\nabla} Y - \underset{Y}{\nabla} X - [X,Y] \tag{24}$$

Substituting the definitions (23) and (11) in this equation gives

$$S(X,Y)^i = (\Gamma^i_{jk} - \Gamma^i_{kj}) X^k Y^j$$

Thus S is a bilinear vector-valued map, i.e. a tensor, called the *torsion* of the connection, having components

$$S^i_{jk} = \Gamma^i_{jk} - \Gamma^i_{kj} \tag{25}$$

In our example of §3.3.7 the first connection discussed had zero torsion, while the second had non-zero torsion. We shall use the vanishing of the torsion as one criterion for the 'naturalness' of a connection.*

From now on we shall only use connections with zero torsion.

From (24) and (25) we therefore have now that

$$\Gamma^i_{kj} = \Gamma^i_{jk} \quad \text{and} \quad \underset{X}{\nabla} Y - \underset{Y}{\nabla} X = [X, Y] \tag{26}$$

3.3.10 The metric connection

We now show that the metric can be used to single out a unique connection, by demanding that, if $Y(u)$ and $Z(u)$ are parallely propagated on a curve γ with parameter u, then the 'inner product' $g(Y(u), Z(u))$ is constant. This must be so when γ is the world-line of a freely falling observer, if his freely falling frame, defined by parallely propagating the vectors $\underset{i}{\bar{E}}$, is to satisfy

$$g(\underset{i}{\bar{E}}, \underset{j}{\bar{E}}) = \overset{0}{g}_{kl}\, \delta^k_i\, \delta^l_j \qquad \text{(in freely falling coordinates)}$$

$$= \overset{0}{g}_{ij} = \text{constant}$$

So for parallel propagation on an arbitrary curve we require

$$g_{ij} Y(u)^i Z(u)^j = \text{constant}$$

or, differentiating and using $\mathrm{d}/\mathrm{d}u = (\mathrm{d}x^k/\mathrm{d}u)\partial/\partial x^k = X^k \partial/\partial x^k$, where $X = \dot{\gamma}$ is the tangent vector to γ,

$$X^k g_{ij,k} Y(u)^i Z(u)^j + g_{ij}\left(\frac{\mathrm{d}Y^i}{\mathrm{d}u} Z^j + Y^i \frac{\mathrm{d}Z^j}{\mathrm{d}u}\right) = 0 \tag{27}$$

Now since Y and Z are parallely propagated we have from §3.3.6 that

$$\frac{\mathrm{d}Y^i}{\mathrm{d}u} = -\Gamma^i_{kj} Y^k X^j$$

and similarly for Z. Thus (27) becomes

$$0 = X^k Y^i Z^j g_{ij,k} - g_{ij}(\Gamma^i_{mk} Y^m X^k Z^j + \Gamma^j_{mk} Z^m X^k Y^i)$$

If this is to hold for all X, Y and Z then

$$g_{ij,k} - g_{lj}\Gamma^l_{ik} - g_{il}\Gamma^l_{jk} = 0 \tag{28a}$$

* If we retain S then in a physical theory it must be related to a physical cause, which is usually the microscopic spin of particles. Thus our assumption $S = 0$ is equivalent to taking this spin to have a negligible net effect. See §3.5.4 (iii).

We now solve this for Γ^l_{ik}. To do this, write the equation twice more, permuting the indices

$$g_{jk,i} - g_{lk}\Gamma^l_{\ ji} - g_{jl}\Gamma^l_{\ ki} = 0 \tag{28b}$$

$$g_{ki,j} - g_{li}\Gamma^l_{\ kj} - g_{kl}\Gamma^l_{\ ij} = 0 \tag{28c}$$

Now form (28*a*) − (28*b*) − (28*c*) and use the zero-torsion condition $\Gamma^l_{\ jk} = \Gamma^l_{\ kj}$ (equation (26)) and the symmetry $g_{il} = g_{li}$ to give

$$g_{ij,k} - g_{jk,i} - g_{ki,j} + 2\Gamma^l_{\ ij} g_{lk} = 0$$

Rearranging and multiplying by the inverse metric g^{km} gives

$$\Gamma^m_{\ ij} = \tfrac{1}{2} g^{km}(-g_{ij,k} + g_{jk,i} + g_{ki,j}) = : \left\{ {m \atop ij} \right\} \tag{29}$$

where we introduce the abbreviation $\left\{ {m \atop ij} \right\}$ for this combination of metric and derivatives. This is called the *Christoffel symbol* of the second kind. The Christoffel symbol of the first kind is defined by

$$[ij, k] = \tfrac{1}{2}(-g_{ij,k} + g_{jk,i} + g_{ki,j})$$

so that

$$\left\{ {m \atop ij} \right\} = g^{km}[ij, k]$$

Any connection for which parallely propagated vectors have constant inner products is called a metric connection. We have just proved that there is a unique torsion-free metric connection, given by (29).

3.3.11 Timelike, spacelike and null geodesics

The tangent vector to a geodesic γ is parallely propagated along the geodesic (by equation (13)) and so for a metric connection

$$g(\dot{\gamma}, \dot{\gamma}) = k, \text{ constant}$$

The value of this constant can be altered by reparametrizing the geodesic (§ 3.3.5). Putting $\tilde{\gamma}(u') := \gamma(au' + b)$ gives $\dot{\tilde{\gamma}} = a\dot{\gamma}$, and so

$$g(\dot{\tilde{\gamma}}, \dot{\tilde{\gamma}}) = a^2 k$$

We can now distinguish three possibilities.

(i) $k > 0$. $\dot{\gamma}$ is spacelike everywhere and there exists a parametrization for which $g(\dot{\gamma}, \dot{\gamma}) = 1$.

(ii) $k = 0$. $\dot{\gamma}$ is null everywhere—a null geodesic.

(iii) $k < 0$. $\dot{\gamma}$ is timelike everywhere and there exists a parametrization for which $g(\dot{\gamma}, \dot{\gamma}) = -1$. A parameter that does this is called a proper time (as defined in § 1.8.1) for the timelike geodesic.

In case (iii) only, γ is a possible world-line for a freely falling observer. Then the time coordinate of freely falling coordinates is a proper time parameter, because if it is used to parametrize γ we have $\dot{\gamma} = \underset{0}{\bar{E}}$ and so $g(\dot{\gamma}, \dot{\gamma}) = \overset{0}{g}_{ij}\delta^i_0\delta^i_0 = \overset{0}{g}_{00} = -1$.

By extension of this idea, we take the timelike geodesics, and only these, as the possible world-lines for freely moving particles, corresponding to the timelike straight lines of special relativity.

The null geodesics are taken as the paths taken by light (more precisely, the wave-fronts of electromagnetic waves are traced out by null geodesics). This corresponds to the special relativity result (§ 2.3.3) that plane electromagnetic waves propagate in the direction of a constant null vector. Indeed, a general relativistic calculation shows that, in the limit as the wavelength of the wave becomes small compared to the overall spatial dimensions of any problem, at any point on a null geodesic defining the path of a wave-front of light the tangent vector to the geodesic corresponds exactly to the null vector k in § 2.3.3, which gives the frequency of the waves.

3.3.12 The construction of a special freely falling coordinate system

Suppose we are given a metric connection on space–time.

Let γ be a timelike geodesic parametrized by proper time passing through an event x_0. We first construct a freely falling frame on γ and then corresponding coordinates.

Putting $\underset{0}{\bar{E}} = \dot{\gamma}$, then an elementary argument in linear algebra shows that we can choose three other vectors $\underset{1}{\bar{E}}, \underset{2}{\bar{E}}, \underset{3}{\bar{E}}$ in $T_{x_0}(M)$ so that $g(\underset{i}{\bar{E}}, \underset{j}{\bar{E}}) = \overset{0}{g}_{ij}$. If we parallely propagate these along γ with a metric connection we shall obtain a freely falling frame with $g(\underset{i}{\bar{E}}, \underset{j}{\bar{E}}) = \overset{0}{g}_{ij}$ everywhere on γ.

Now we construct coordinates having $\{\underset{i}{\bar{E}}\}$ as a coordinate basis. It can be shown that there is a neighbourhood of γ in which each point y lies on a spacelike geodesic γ_y, unique apart from reparametrization, that starts at a point $\gamma(s_y)$ on $\gamma(\gamma(s_y) = \gamma_y(0))$ and finishes at $y = \gamma_y(u)$ for some u, and that also has $\dot{\gamma}_y(0) = \sum_{\alpha=1}^{3} z^\alpha \underset{\alpha}{\bar{E}}$ for some numbers z^1, z^2, z^3 (see Figure 7).

The freely falling coordinates $\bar{\mathbf{x}}$ are now defined by

$$\bar{x}^0(y) := s_y$$

$$\bar{x}^\alpha(y) := uz^\alpha \qquad (\alpha = 1, 2, 3)$$

Note that this is independent of the parametrization of γ_y; if we define $\tilde{\gamma}_y(v') := \gamma_y(av)$ then $\tilde{\gamma}_y(\tilde{u}) = y$, where $\tilde{u} = u/a$, and

$\dot{\gamma}_y(0) = \sum_{\alpha=1}^{3} \tilde{z}^\alpha \underset{\alpha}{\bar{E}}$ where $\tilde{z}^\alpha = a z^\alpha$; thus $\tilde{u}\,\tilde{z}^\alpha = u z^\alpha$ and the two parametrizations define the same coordinates.

It is clear from the construction that the frame $\{\underset{i}{\bar{E}}\}$ is now composed of the basis vectors of this coordinate system. The important feature of this particular construction is the following:

Proposition The coordinates $\bar{\mathbf{x}}$ constructed above are such that on γ we have $\bar{\Gamma}^i_{jk} = 0$.

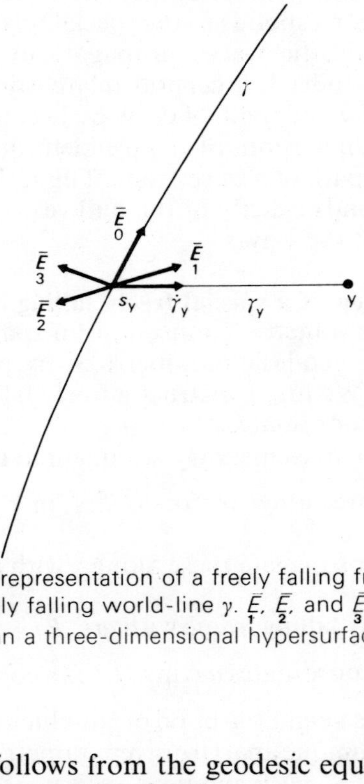

Figure 7. Schematic representation of a freely falling frame $\underset{\alpha}{\bar{E}}$ at a point with parameter s_y on a freely falling world-line γ. $\underset{1}{\bar{E}}$, $\underset{2}{\bar{E}}$ and $\underset{3}{\bar{E}}$ are actually mutually perpendicular and span a three-dimensional hypersurface containing the vector $\dot{\gamma}_y$ tangent to γ_y.

Proof The result follows from the geodesic equations for γ_y and the parallel propagation equation for $\underset{i}{E}$.

First note that each point on γ_y has the same set of values for (z^1, z^2, z^3), different points having different values of u. Thus the equation of γ_y is

$$\bar{x}^\alpha(\gamma_y(u)) = u z^\alpha$$
$$\bar{x}^0(\gamma_y(u)) = s_y$$

And so $\mathrm{d}x^\alpha/\mathrm{d}u = z^\alpha$, $\mathrm{d}^2x^\alpha/\mathrm{d}u^2 = 0$, $\mathrm{d}x^0/\mathrm{d}u = 0$.

Inserting this in the geodesic equation (14) gives

$$\sum_{\alpha,\beta=1}^{3} \bar{\Gamma}^i_{\alpha\beta} z^\alpha z^\beta = 0 \tag{30}$$

valid on γ_y, and in particular at $\gamma(s_y)$. For each set of values of (z^1, z^2, z^3) there is a geodesic of the form $\gamma_{y'}$ through $\gamma(s_y)$ (for some y') and so (30) holds for all values of the z^α, at each point of γ. Since Γ^i_{jk} is symmetric, by Exercise 5 (iii) of § 1.6.4 this implies that

$$\bar{\Gamma}^i_{\alpha\beta} = 0 \qquad (\alpha, \beta = 1, 2, 3) \text{ on } \gamma$$

Next, the equation for parallel propagation (§ 3.3.6) applied to $\underset{i}{\bar{E}}$ gives

$$0 = \left(\frac{\mathrm{D}\underset{i}{\bar{E}}}{\mathrm{D}s}\right)^j = \frac{\mathrm{d}}{\mathrm{d}s}\delta^j_i + \bar{\Gamma}^j_{kl}\dot{\bar{\gamma}}^l\delta^k_i = \bar{\Gamma}^j_{io} \tag{31}$$

Thus (30) and (31) together give the result.

The coordinates just constructed are a special case of Fermi–Walker coordinates.

3.3.13 Summary of the model

Space–time M (described in § 3.2.3) is furnished with a metric g (described in § 3.3.1). This singles out a unique torsion-free metric connection, given by $\Gamma^i_{jk} = \left\{ {i \atop jk} \right\}$, (§ 3.3.10). The world-lines of freely falling observers are the timelike geodesics with respect to this metric and connection, and their freely falling frames are obtained by parallel propagation, (§ 3.3.6). Of the many coordinate systems having one of these frames as its coordinate basis on some timelike geodesic there is one, $\bar{\mathbf{x}}$, for which $\bar{g}_{ij} = \overset{0}{g}_{ij}$ and $\bar{\Gamma}^i_{jk} = 0$ on the geodesic. For this coordinate system the covariant derivative *on the geodesic* is simply given by $\underset{X}{\nabla} = \bar{X}^i\partial/\partial\bar{x}^i$ (as in special relativity).

From the transformation law for Γ^i_{jk} (Exercise 4), any coordinates $\mathbf{x}'$ having the same coordinate basis and vanishing second derivatives with respect to $\bar{\mathbf{x}}$ on the geodesic will also have $g'_{ij} = \overset{0}{g}_{ij}$ and $\Gamma'^i_{jk} = 0$. From now on we shall refer to all such coordinates, and only such coordinates, as *freely falling*.

3.3.14 Notation for the metric

To conclude this section, we give a standard way of writing the metric that makes coordinate transformations and similar manipulations very easy. We use the basis covectors $\mathrm{d}x^0, \ldots, \mathrm{d}x^3$ introduced in § 1.9.1: the arguments there carry over precisely to general relativity, showing that, for each $i = 0, 1, 2, 3$, $\mathrm{d}x^i$ is a covector field throughout the region where the coordinates are defined, and $\mathrm{d}x^i(x)$ (a covector in $T^*_x(M)$) has components δ^i_j. The tensor product (§ 1.7.1) of two of these covectors, $\mathrm{d}x^i \otimes \mathrm{d}x^k$, is thus a tensor of type (0, 2) with components $(\mathrm{d}x^i \otimes \mathrm{d}x^k)_{jl} = \delta^i_j \delta^k_l$. These tensors form a basis for $T^{(0,\,2)}_x(M)$ at each point x, and in particular we can express the metric as

$$g = g_{ik} \mathrm{d}x^i \otimes \mathrm{d}x^k$$

(proof by taking components of both sides).

When the summation implied here is written out in full in a particular example, it is usual to simplify it further by introducing the *symmetrized tensor product* $\omega\nu$ of two covectors, ω and ν defined by

$$\omega\nu := \tfrac{1}{2}(\omega \otimes \nu + \nu \otimes \omega)$$

Then we can write

$$g = g_{ij} \mathrm{d}x^i \otimes \mathrm{d}x = \tfrac{1}{2}(g_{ij} + g_{ji}) \mathrm{d}x^i \otimes \mathrm{d}x^j$$
$$= g_{ij} \mathrm{d}x^i \mathrm{d}x^j$$

This form is particularly useful when changing coordinates. For, if we express $\mathbf{x}$ in terms of new coordinates $\mathbf{x}'$, then we obtain the form of the metric in the new coordinates simply by substituting

$$\mathrm{d}x^i = (\partial x^i / \partial x'^k)\, \mathrm{d}x'^k$$

(as shown in § 1.9.1) in the above expression. We shall encounter an example of this technique in § 5.1.3.

Finally, the product $\omega\omega$ is usually written ω^2. Thus $(\mathrm{d}x^0)^2$ means $\mathrm{d}x^0 \otimes \mathrm{d}x^0$, and so on.

3.4 Curvature

3.4.1 Formal properties of ∇

So far we have defined the covariant derivative only for functions (when the covariant derivative along X is simply the directional derivative) and for vectors. We now extend it to all tensors in such a way that it obeys the same algebraic relations as the directional derivative; it is quite easy to show that there is only one way of doing this.

Definition If X is a vector in $T_x(M)$ and T is a tensor field of type (n, m) defined in some neighbourhood of x, then we define $\underset{X}{\nabla} T$ to be the tensor of type (n, m) at x with components

$$(\underset{X}{\nabla} T)^{ij\cdots l}{}_{pq\cdots u} = X^k T^{ij\cdots l}{}_{pq\cdots u;k}$$

where

$$T^{ij\cdots l}{}_{pq\cdots u;k} = T^{ij\cdots l}{}_{pq\cdots u,k} + \Gamma^{i}_{mk} T^{mj\cdots l}{}_{pq\cdots u}$$
$$+ \Gamma^{j}_{mk} T^{im\cdots l}{}_{pq\cdots u} + \ldots + \Gamma^{l}_{mk} T^{ij\cdots m}{}_{pq\cdots u} - \Gamma^{m}_{pk} T^{ij\cdots l}{}_{mq\cdots u}$$
$$- \Gamma^{m}_{qk} T^{ij\cdots l}{}_{pm\cdots u} - \ldots - \Gamma^{m}_{uk} T^{ij\cdots l}{}_{pq\cdots m} \quad (32)$$

(i.e. the partial derivative, with one '$+\Gamma$' factor added for each upper index and one '$-\Gamma$' for each lower index).

We can now show that the covariant derivative of a tensor product is found by the usual rule for the derivative of a product (the so-called Leibnitz' rule), and that it commutes with contraction. To formulate this last, define $C^{(r)}_{(s)}T$ to be the contraction of T (§ 1.7.3) with respect to its rth upper index and sth lower index, that is

$$(C^{(r)}_{(s)}T)^{ij\cdots}{}_{pq\cdots} := T^{ij\cdots \overset{(r)}{m} \cdots}{}_{pq\cdots \underset{(s)}{m} \cdots}$$

Proposition 1
(a) For any tensor fields U and V in a neighbourhood of x and any X in $T_x(M)$ we have

$$\underset{X}{\nabla}(U \otimes V) = (\underset{X}{\nabla}U) \otimes V + U \otimes \underset{X}{\nabla}V \quad (33)$$

(b) For any tensor field of type (m, n) with $m \geq r$, $n \geq s$ in a neighbourhood of x we have

$$\underset{X}{\nabla}(C^{(r)}_{(s)}T) = C^{(r)}_{(s)}\underset{X}{\nabla}T \quad (34)$$

Proof (a) follows immediately from the definition (32); (b) is almost as trivial: on contracting a covariant derivative the Γ associated with the upper contracted index cancels the $-\Gamma$ associated with the lower one, leaving the derivative of the contracted tensor. (The details are left to the reader.)

Corollary If ω is a covector field and Y a vector field, then

$$\underset{X}{\nabla}(\omega(Y)) = (\underset{X}{\nabla}\omega)(Y) + \omega(\underset{X}{\nabla}Y) \quad (35)$$

Proof We simply note that $\omega(Y) = \omega_i Y^i = C^{(1)}_{(1)}(\omega \otimes Y)$, and then apply both parts of the proposition.
(A similar result holds for $k(Y, Z)$, k of type $(0, 2)$, etc.)

Proposition 2

$$\underset{X}{\nabla}g = 0 \qquad \text{for all } X \quad (36)$$

Proof

$$(\underset{X}{\nabla} g)_{ij} = X^k(g_{ij,k} - \Gamma^m_{ik} g_{mj} - \Gamma^m_{jk} g_{im}) \qquad \text{from (32)}$$
$$= 0 \qquad \text{from (28)}$$

The equivalence of this result to equation (28) shows that $\underset{X}{\nabla} g = 0$ is an alternative formulation of the condition that the connection be metric. It also has the important consequence for the raising/lowering convention that the operations of covariant differentiation and raising or lowering commute:

$$(T^{i\ k\cdots}_{\ j})_{jm} = (g_{jl} T^{ilk\cdots})_{;m} = g_{jl}(T^{ilk\cdots}{}_{;m}) \qquad (36')$$

Thus there is no ambiguity in continuing to use the convention along with the $_{;k}$ notation for the covariant derivative.

3.4.2 Definition of the Riemann tensor Riem

In § 3.3.9 we investigated the effect of reversing the order of covariant derivatives acting on a function; we now repeat this with the derivatives acting on a vector field.

Let X, Y and Z be vector fields. Then

$$(\underset{X}{\nabla}\underset{Y}{\nabla} Z - \underset{Y}{\nabla}\underset{X}{\nabla} Z)^i = X^j(Y^k Z^i{}_{;k})_{;j} - Y^j(X^k Z^i{}_{;k})_{;j}$$
$$= X^j Y^k{}_{;j} Z^i{}_{;k} - Y^j X^k{}_{;j} Z^i{}_{;k} + (X^j Y^k - Y^j X^k) Z^i{}_{;kj}$$
$$= [X, Y]^k Z^i{}_{;k} + X^j Y^k Z^i{}_{;kj} - Y^k X^j Z^i{}_{;jk}$$

(using (26) for the first term and interchanging the dummy suffices j and k in the third term), i.e.

$$(\underset{X}{\nabla}\underset{Y}{\nabla} Z - \underset{Y}{\nabla}\underset{X}{\nabla} Z)^i = (\underset{[X,\,Y]}{\nabla} Z)^i + X^j Y^k (Z^i{}_{;kj} - Z^i{}_{;jk}) \qquad (37)$$

The second term can be worked out as follows, using the expression (32) for the covariant derivative.

$$Z^i{}_{;kj} = (Z^i{}_{,k} + \Gamma^i_{lk} Z^l)_{;j}$$
$$= Z^i{}_{,kj} + \Gamma^i_{lk,j} Z^l + \Gamma^i_{lk} Z^l{}_{,j}$$
$$+ \Gamma^i_{nj}(Z^n{}_{,k} + \Gamma^n_{lk} Z^l) - \Gamma^m_{kj}(Z^i{}_{,m} + \Gamma^i_{lm} Z^l)$$

Subtracting the same expression with k and j interchanged, and using the symmetries $Z^i{}_{,kj} = Z^i{}_{,jk}$ and $\Gamma^m_{kj} = \Gamma^m_{jk}$ (26) gives

$$Z^i{}_{;kj} - Z^i{}_{;jk} = (\Gamma^i_{lk,j} - \Gamma^i_{lj,k} + \Gamma^i_{nj}\Gamma^n_{lk} - \Gamma^i_{nk}\Gamma^n_{lj}) Z^l \qquad (38)$$

Substituting this in (37) then gives

$$\underset{X}{\nabla}\underset{Y}{\nabla} Z - \underset{Y}{\nabla}\underset{X}{\nabla} Z = \underset{[X,\,Y]}{\nabla} Z + \mathrm{Riem}(Z, X, Y) \qquad (39)$$

where we have written 'Riem' to denote the Γ-terms on the right-hand-side of (38), viz.

$$(\mathrm{Riem}(Z,X,Y))^i := R^i{}_{ljk} Z^l X^j Y^k \tag{40}$$

$$R^i{}_{ljk} := \Gamma^i_{lk,j} - \Gamma^i_{lj,k} + \Gamma^i_{nj}\Gamma^n_{lk} - \Gamma^i_{nk}\Gamma^n_{lj} \tag{41}$$

It is clear from (39) and (40) that Riem is a trilinear map from vectors to vectors, and so a tensor, called the *Riemann* or *curvature tensor*.

It can be shown that, if the Riemann tensor is zero in a region of space–time that is toplogically well-behaved, then the result of parallely transporting a vector from x to y is independent of the path taken. Thus the fact that in general Riem $\neq 0$ is responsible for the path-dependence of parallel transport, and so for the fact that it is impossible to set up a global freely falling coordinate system in general relativity. In this sense, Riem expresses the departure of the space–time from the Minkowski form.

Notation When using the index notation it is customary to denote the components of Riem, and also the components of a tensor of type (0, 2) and a scalar derived from it, all by the symbol R (the number of indices, if any, serving to distinguish the three different objects). Thus we shall define the following:

(i) the Ricci* tensor, Ric, with components $(\mathrm{Ric})_{lk} \equiv R_{lk} := R^i{}_{lik}$ (42)

(ii) the Ricci scalar, or curvature scalar $R := R^k{}_k = R^{ik}{}_{ik}$ (43)

It should be noted that there is no general agreement on the sign of the Riemann tensor (many authors calling $-R^i{}_{jkl}$ what we call $R^i{}_{jkl}$). Also it is more usual to write Riem(Z,X,Y) as $R(X,Y)Z$.

3.4.3 Properties of Riem

The curvature tensor satisfies various identities as a consequence of its definition. To prove these it is convenient to work first in a freely falling coordinate system, and then transfer to an arbitrary system. The identities, being expressed in terms of tensors, will retain the same form in a general coordinate system. We make use of the fact that in freely falling coordinates $\bar{\Gamma}^i_{jk} = 0$ (§ 3.3.13) and hence, from (28), $\bar{g}_{ij,k} = 0$, both holding at any point on the geodesic from which the coordinates are defined, but not elsewhere.

So, choosing an arbitrary event x_0 and drawing any timelike geodesic through it, we can construct a freely falling coordinate system. Then at x_0, *in these coordinates,*

$$\begin{aligned}\bar{R}^i{}_{ljk} &= \bar{\Gamma}^i_{lk,j} - \bar{\Gamma}^i_{lj,k} \\ &= \tfrac{1}{2}\bar{g}^{im}(-\bar{g}_{lk,mj} + \bar{g}_{km,lj} + \bar{g}_{ml,kj} + \bar{g}_{lj,mk} - \bar{g}_{jm,lk} - \bar{g}_{ml,jk})\end{aligned}$$

*Other authors define $R_{lk} := R^i{}_{lki}$

and so

$$\bar{R}_{mljk} = \tfrac{1}{2}(-\bar{g}_{lk,mj} + \bar{g}_{km,lj} + \bar{g}_{lj,mk} - \bar{g}_{jm,lk})$$

Hence we have immediately that

$$R_{mljk} = R_{jkml} \tag{44}$$

$$R_{mljk} + R_{mjkl} + R_{mklj} = 0 \tag{45}$$

and

$$R_{mljk} = -R_{mlkj} \tag{46}$$

(the last being equally obvious from the definition of Riem).
From (44) and (46) we have

$$R_{mljk} = -R_{lmjk} \tag{47}$$

There is also an identity involving the covariant derivative of Riem. At x_0, in freely falling coordinates

$$\bar{R}^i{}_{ljk;n} = \bar{R}^i{}_{ljk,n} = \bar{\Gamma}^i_{lk,jn} - \bar{\Gamma}^i_{lj,kn}$$

and thus

$$R^i{}_{ljk;n} + R^i{}_{lkn;j} + R^i{}_{lnj;k} = 0 \tag{48}$$

This is called *Bianchi's identity.*

As we remarked above, (44)–(48) are valid, by application of the transformation laws, in any coordinate system and, since x_0 was initially chosen arbitrarily, at any point.

3.5 Gravitation

An observer in free fall feels no gravitational forces only in the limit as the region being considered becomes very small (compared to a typical length-scale for the variation of the gravitational field). He can discover that he is under the influence of gravity, and not in empty space (apart from himself!) by noting that an object a short distance away, where the gravitational field is in a slightly different direction, has a slight acceleration relative to himself. We shall now investigate this acceleration within the model we have developed.

3.5.1 Geodesic deviation

Let γ_0 be the world-line of the observer considered in the preceding section and γ_1 the world-line of a nearby body, both lines being geodesics parametrized by proper time. To investigate the limiting behaviour as the separation becomes small, interpolate a whole family of geodesics γ_t between γ_0 and γ_1, where t varies between 0 and 1, doing this in any way provided that the function $\gamma_t(s)$ for fixed s varies smoothly with t, and hence describes a curve in space–time (Figure 8). We expect that the limiting behaviour as the separation of the geodesics becomes small, i.e. as $t \to 0$, can be expressed in terms of the tangent vector Y to this curve at $t = 0$. Indeed, the separation of $\gamma_{\delta t}$ from γ_0, for small δt, is given by

$$\begin{aligned} \delta x^i := {} & x^i(\gamma_{\delta t}(s)) - x^i(\gamma_0(s)) \\ = {} & \delta t \partial x^i(\gamma_t(s))/\partial t|_{t=0} + \mathrm{O}(\delta t) \\ = {} & \delta t Y^i(s,0) + \mathrm{O}(\delta t) \end{aligned} \tag{49}$$

since

$$Y^i(s,t) := \frac{\partial x^i(\gamma_t(s))}{\partial t} \tag{50}$$

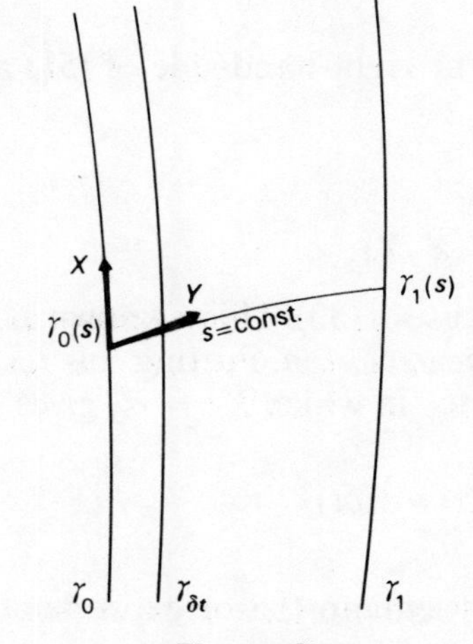

Figure 8

The acceleration of the nearby particles relative to γ_0 is given by the second derivative of $\delta \bar{x}^i$ with respect to proper time s, evaluated in freely falling coordinates $\bar{\mathbf{x}}$. Using these coordinates in (49) and expressing Y in a freely falling frame as $Y = \bar{Y}^i \underset{i}{\bar{E}}$ we see that the acceleration of $\gamma_{\delta t}$ relative to γ_0 is given by

$$\frac{\mathrm{d}^2 \delta \bar{x}^i}{\mathrm{d}s^2} \approx \delta t \frac{\mathrm{d}^2 \bar{Y}^i}{\mathrm{d}s^2} = \delta t \left(\overline{\frac{\mathrm{D}^2 \mathrm{Y}}{\mathrm{D}s^2}} \right)^i = \delta t (\overline{\underset{X}{\nabla} \underset{X}{\nabla} Y})^i \tag{51}$$

to first order in δt, since covariant and ordinary derivatives are the same along γ; as usual $X = \dot{\gamma}$ denotes the tangent vector to the geodesics, i.e. we set

$$X^i = \frac{\partial x^i(\gamma_t(s))}{\partial s} \tag{52}$$

Now from (26)

$$(\underset{X}{\nabla} Y - \underset{Y}{\nabla} X)^i = [X, Y]^i = X^j(\partial/\partial x^j)(\partial x^i/\partial t) - Y^j(\partial/\partial x^j)(\partial x^i/\partial s)$$
$$= \partial^2 x^i/\partial s \partial t - \partial^2 x^i/\partial t \partial s = 0 \tag{53}$$

from the definitions (50) and (52) of Y and X and the fact that $X^j \partial/\partial x^j = \partial/\partial s$ and similarly for Y and t. (Strictly speaking, this and subsequent formulae quoted only apply to vector fields, whereas X and Y are defined only on the family of geodesics. Thus we must arbitrarily extend them to form fields. In fact, the domain of X and Y need only be such that all the expressions involved are well defined if one goes back to the original definitions.)

Thus we can write the right-hand-side of (51) as

$$\frac{\mathrm{D}^2 Y}{\mathrm{D}s^2} = \underset{X}{\nabla}\underset{Y}{\nabla} X$$
$$= \mathrm{Riem}(X, X, Y) \tag{54}$$

from (39) and a second use of (53). This is known as *Jacobi's equation* or the *equation of geodesic deviation*. Putting this result in (51) and using freely falling coordinates in which $\bar{X}^i = \delta^i_0$ gives, with (49),

$$\frac{\mathrm{d}^2 \delta \bar{x}^i}{\mathrm{d}s^2} = \bar{R}^i{}_{00j} \delta \bar{x}^j + \mathrm{o}(\delta t) \tag{55}$$

So we see that the Riemann tensor expresses directly the effect of gravity in a freely falling frame.

3.5.2 The source of gravitation

We now compare equation (55) with Newtonian theory , with the aim (as with special relativity in chapter 2) of finding an acceptable general relativistic law that will approximate to the Newtonian one under the conditions in which the latter has been experimentally verified—for slowly moving matter and moderate pressures. We shall work in freely falling coordinates and, as in Newtonian theory, treat time and space in different ways; for now, we forget about the geometric significance of vectors and suspend our usual conventions about positions of indices.

In Newtonian gravitation theory there is a gravitational potential ϕ which produces an acceleration $\mathbf{f} = -\nabla\phi$; in Cartesian components

$f_\alpha = -\phi_{,\alpha}$ ($\alpha = 1, 2, 3$). Thus the relative acceleration of two nearby particles at $\mathbf{r}$ and $\mathbf{r}+\delta\mathbf{r}$ is

$$\frac{d^2}{dt^2}\delta r_\alpha = \delta f_\alpha = f_\alpha(\mathbf{r}+\delta\mathbf{r}) - f_\alpha(\mathbf{r})$$
$$= \sum_\beta f_{\alpha,\beta}\delta r_\beta + o(\delta\mathbf{r})$$

We can compare this with (55), in which we can identify s, the proper time, with the Newtonian time (provided the velocities involved are small compared with unity, the velocity of light, cf. equation (2) of chapter 2). To make the comparison closer, note that $\bar{R}^i{}_{000} = 0$ from (46), so that the spatial components of (55) can be written

$$\frac{d^2\delta\bar{x}^\alpha}{ds^2} = \sum_{\beta=1}^{3} \bar{R}^\alpha{}_{00\beta}\delta\bar{x}^\beta + o(\delta t)$$

Thus (in freely falling coordinates)

$$\bar{R}^\alpha{}_{00\beta} \quad \text{corresponds to} \quad f_{\alpha,\beta}\,(= -\phi_{,\alpha\beta}) \tag{56}$$

In Newtonian theory ϕ is related to the density ρ by

$$\nabla^2\phi \equiv \sum_{\alpha=1}^{3} \phi_{,\alpha\alpha} = 4\pi G\rho \tag{57}$$

where G is the gravitational constant.

To find the corresponding equation in general relativity, we use the result discussed in chapter 2, that the density appears as the component T_{00} of a tensor in special relativity. Strictly speaking, this component is the energy density, but in special relativity mass is a form of energy and makes up most of the energy of matter, if it is moving slowly. In view of (56), if we are to have agreement with the Newtonian equation (57) we must have, in the Newtonian limit

$$\bar{R}^i{}_{00i} \equiv \sum_{\alpha=1}^{3} \bar{R}^\alpha{}_{00\alpha} \approx -4\pi G\bar{T}_{00}$$

or, using (46), (47) and the definition (43) of R_{ik}

$$\bar{R}_{00} \approx 4\pi G\bar{T}_{00} \tag{58}$$

This looks like the 00-component of a tensor equation $R_{ij} = 4\pi G T_{ij}$, and this was indeed the first equation thought of by Einstein. He rejected it for reasons that were invalid; the true reason why this law will not work will be given later (§3.5.3). It took Einstein three years to reach the following modification of the law:

$$R_{ij} - \tfrac{1}{2}Rg_{ij} = 8\pi G T_{ij} \tag{59}$$

To see why this is equivalent to (58) in the Newtonian limit, we note that, in the natural units we are using here, $\bar{T}_{00}$ is the dominant term of $\bar{T}_{ij}$ for matter in normal states of stress and pressure. (In terms of conventional units, in which the velocity of light is c, $\bar{T}_{11}$, for instance, is p/c^2, where p is the pressure in the x-direction, which is usually very much less than $\bar{T}_{00} = \rho$.) If we neglect all the terms in $\bar{\mathbf{T}}$ apart from $\bar{T}_{00}$, therefore, we obtain

$$\bar{T}^i{}_i \approx \bar{T}^0{}_0 = -\bar{T}_{00} \quad (= -\rho) \tag{60}$$

We now solve (59) for R by first contracting to give

$$R^i{}_i - \tfrac{1}{2}R\delta^i_i = -R = 8\pi G T^i{}_i$$

so that, eliminating R from (59)

$$R_{ij} = 8\pi G(T_{ij} - \tfrac{1}{2}T^k{}_k g_{ij}) \tag{61}$$

Hence, putting $i = j = 0$, the approximation (60) gives (58) as required.

The equation (59) is therefore adopted as giving the connection between the gravitational field and the energy–momentum tensor as its source: it is usually called Einstein's equation, and is the simplest physically acceptable relationship within the general model we have developed.

3.5.3 The contracted Bianchi identities

If we raise the index l in (48) and change $k \to i$, $n \to l$ we obtain (using (36′))

$$0 = R^{il}{}_{ji;l} + R^{il}{}_{il;j} + R^{il}{}_{lj;i}$$

or
$$0 = -R^l{}_{j;l} + R^l{}_{l;j} - R^i{}_{j;i} = (-2R^l{}_j + R^k{}_k\delta^l{}_j)_{;l} \tag{62}$$

and hence Einstein's equation (59) then gives

$$T^l{}_{j;l} = 0$$

This is the equation that we have already found in special relativity (equation (31) of chapter 2), with partial derivatives rather than covariant, as expressing the law of conservation of momentum. So this law governing the motion of the matter is, in general relativity, a consequence of the Einstein equation linking matter to the gravitational field.

The relation (62) also provides the reason for rejecting the simpler equation

$$R_{ij} = 4\pi G T_{ij}$$

because, if we put this in (62) we obtain

$$(T^l{}_j - \tfrac{1}{2}T^k{}_k\delta^l_j)_{;l} = 0$$

If we now demand, on the basis of the special relativity result (31) of chapter 2, that $T^l{}_{j;l} = 0$ is a separate condition, then the above equation gives $T^k{}_{k,l} = 0$. In other words, the only possible distributions of matter are those in which $T^k{}_k$ is constant, or if (60) is a good approximation, where the density is constant throughout space–time! The only way to avoid this absurd restriction is to use the form (59) which makes the law $T^l{}_{j;l} = 0$ a consequence of the equations, rather than a separate restriction to be imposed.

3.5.4 Some alternatives to general relativity

Apart from radical alternatives that do not use the differentiable manifold model for space–time (§3.1.3) at all, the following are some of the most popular competitors against general relativity.

(i) The cosmological constant. Later in his life Einstein proposed the following modification of (59):

$$R_{ij} - \tfrac{1}{2}Rg_{ij} + \Lambda g_{ij} = 8\pi G T_{ij}$$

where Λ is a constant, called the cosmological constant. While this makes significant alterations to possible cosmological solutions of the equations, there is little basic difference from a mathematical point of view, since the additional terms could be regarded as an additional contribution of $-\Lambda g$ to the total T (though such an addition to T could not arise from the inclusion of any ordinary sort of matter).

(ii) The Hoyle–Narlikar theory. Although this results in equations that are very similar to those of conventional general relativity, the basic principles are different in that emphasis is laid on direct interactions of distant particles on each other, rather than on particles producing a field in their vicinity that then propagates to distant particles. This allows the possibility of particles being spontaneously created (or annihilated), this being governed by an effective field—the C-field—that enters the field equations for Ric so as to mimic a negative pressure distributed throughout space–time. The ideas are explained in their book *Action at a distance in physics and cosmology* (F. Hoyle and J. V. Narlikar, Freeman, New York, 1974).

(iii) Torsion theories. The inclusion of torsion (§3.3.9) was investigated by Einstein and Cartan, using a different formalism from the one adopted here, and independently developed from different principles by Kibble and Sciama. There is now general agreement that, if torsion is introduced, it is related to the spin of matter, though there is some variation in the details. Most of the accounts are rather technical, but a readable survey is presented by F. Hehl in *General relativity and gravitation*, **4**, p. 333, 1973.

(iv) Scalar–tensor theories. In these gravitation is described not just by the metric g and the quantities derived from it, but by g and a scalar field ϕ. The most popular theory of this kind, due to Jordan, Brans and Dicke, uses field equations of the form

$$R_{ij}-\tfrac{1}{2}Rg_{ij}=\frac{8\pi G}{\phi}(T_{ij}+T^{\phi}_{ij})$$

where T^{ϕ}_{ij} is a tensor depending on ϕ and interpreted as the energy–momentum of the ϕ-field. ϕ is in turn determined by the matter T via a wave equation with T as source.
An account of this and other theories, involving other structures in addition to g, is given by W.-T. Ni, *Astrophysical Journal*, **176**, pp. 769–96, 1972.

Exercises

1 If M is Minkowski space (i.e. $\mathbb{R}^4$ with metric $g_{ij}=\overset{0}{g}_{ij}$) show that under the transformation to the spherical polar coordinates x' given by

$$x^0 = x'^0$$

$$x^1 = x'^1 \sin(x'^2)\cos(x'^3)$$

$$x^2 = x'^1 \sin(x'^2)\sin(x'^3)$$

$$x^3 = x'^1 \cos(x'^2)$$

the metric components become

$$g'_{00}=-1,\quad g'_{11}=1,\quad g'_{22}=(x'^1)^2,\quad g'_{33}=(x'^1)^2\sin^2(x'^2)$$

with the rest zero.

2 Let S be the unit 2-sphere

$$S=\{(x^1,x^2,x^3)\colon (x^1)^2+(x^2)^2+(x^3)^2=1\}$$

regarded as a subset of $\mathbb{R}^3$. Coordinates in $\mathbb{R}^3$ will be denoted by x^α ($\alpha = 1,2,3$) and x'^A ($A = 1,2$) denote the polar coordinates $x'^1 \equiv \theta$, $x'^2 \equiv \phi$; the coordinates are related by $x^1=\sin\theta\cos\phi$, $x^2=\sin\theta\sin\phi$, $x^3=\cos\theta$.
(i) If $\gamma\colon(0,1)\to S\subset\mathbb{R}^3$ is a differentiable curve, calculate the components $X^\alpha := dx^\alpha(\gamma(s))/ds$ of $\dot\gamma$ in terms of the components $X'^A := dx'^A(\gamma(s))/ds$.

[Note that any vector in $T(S)$ can be given components in this way as an element of $T(\mathbb{R}^3)$ (i.e. $T(S) \subset T(\mathbb{R}^3)$).]
(ii) If ${}^{(3)}g$ is the metric on $\mathbb{R}^3$, specified by

$$ {}^{(3)}g(X,Y) = (X^1)^2 + (X^2)^2 + (X^3)^2 $$

show that the metric g on S with components (in the coordinates $\mathbf{x}'$)

$$ g'_{11} = 1, \quad g'_{22} = \sin^2\theta, \quad g'_{12} = g'_{21} = 0 $$

satisfies $g(X,Y) = {}^{(3)}g(X,Y)$ for $X, Y \in T(S) \subset T(\mathbb{R}^3)$.

3 Show that the symbols $\left\{ {A \atop BC} \right\}$ for the 2-sphere with metric

$$ g = (\mathrm{d}x^1)^2 + (\sin(x^1))^2 (\mathrm{d}x^2)^2 $$

(as in Exercise 2(ii)) give the connection Γ^A_{BC} of §3.3.7. (The indices A, B, C range over 1, 2.)

4 By calculating $\underset{X}{\nabla} Y$ in two coordinate systems $\mathbf{x}'$ and $\mathbf{x}$, and equating

$$ (\underset{X}{\nabla} Y)^i = \frac{\partial x^i}{\partial x'^j} (\underset{X}{\nabla} Y)' \qquad \text{(for all } X, Y) $$

deduce from (5) the transformation law for Γ as

$$ \Gamma'^i_{mj} = \frac{\partial x^p}{\partial x'^j} \frac{\partial x^q}{\partial x'^m} \left(\frac{\partial x'^i}{\partial x^l} \Gamma^l_{qp} - \frac{\partial^2 x'^i}{\partial x^p \partial x^q} \right) $$

5 Let γ be a differentiable curve from $x := \gamma(0)$ to $y := \gamma(1)$ and put $\mathbf{x}(s) := \mathbf{x}(\gamma(s))$, $\dot{\mathbf{x}} := \mathrm{d}\mathbf{x}/\mathrm{d}s$.
(i) Let

$$ I(x, y, \gamma) = \int_0^1 f(\mathbf{x}(s), \dot{\mathbf{x}}(s)) \mathrm{d}s $$

where

$$ f(\mathbf{x}(s), \dot{\mathbf{x}}(s)) := g_{ij}(\mathbf{x}(s)) \dot{\mathbf{x}}^i \dot{\mathbf{x}}^j $$
$$ (= g(\gamma(s))(\dot{\gamma}(s), \dot{\gamma}(s))) $$

Show that the value of I is stationary for variations of γ with x and y fixed when γ is a geodesic.
(ii) The same question as part (i) but with

$$ f(\mathbf{x}(s), \dot{\mathbf{x}}(s)) := (g_{ij}(\mathbf{x}(s)) \dot{x}^i \dot{x}^j)^{\frac{1}{2}} $$

(It is a result of the calculus of variations that I is stationary when the Euler–Lagrange equations

$$ \frac{\mathrm{d}}{\mathrm{d}s} \frac{\partial f}{\partial \dot{x}^i} = \frac{\partial f}{\partial x^i} $$

are satisfied.)

6 Suppose that g_{ij} is independent of x^1 (i.e. $g_{ij,1} = 0$). Show that, if X is the tangent vector to a geodesic, then the number $\kappa := g_{1j}X^j$ is constant on the geodesic (i.e. $X^l\kappa_{,l} = 0$).

7 A connection Γ^i_{jk} on $\mathbb{R}^4$ (coordinates x^0, x^1, x^2, x^3) has $\Gamma^1_{11} = x^2$ and the other components zero. Investigate the effect of parallely transporting a vector round the edge of the unit square $0 \leq x^1 \leq 1$, $0 \leq x^2 \leq 1$, $x^0 = x^3 = 0$. Deduce that this Γ cannot be a metric connection for any metric.

8 If the members γ_t of a family of geodesics (as in §3.5.1) are parametrized by proper time and Y has components $Y^i = \partial x^i(\gamma_t(s))/\partial t$, prove that $g(\dot{\gamma}_t, Y)$ is constant along the geodesic γ_t.

9 Prove that the Ricci scalar for the 2-sphere (defined as for space–time but with indices ranging over 1, 2), with metric as in Exercise 2(ii), is constant.

4
Weak-field theory

4.1 The theory

Einstein's field equation (59) of chapter 3 can be regarded as a set of non-linear second-order partial differential equations specifying g_{ij} in terms of T_{ij}. The non-linearity of these equations makes them extremely difficult to handle, and only a limited number of solutions (some of which will be discussed in the next chapter) are known. To obtain an idea of the form more general solutions may take the field equation can be replaced by different, linear equations—the weak-field equations—that are approximately the same as Einstein's field equations when the gravitational forces are weak. These linear equations can be solved comparatively easily.

Unfortunately, there is no mathematical reason for supposing that the solutions of the weak-field equations should approximate to the solutions of the full Einstein equations except in very limited cases. This is because, firstly, the solutions of non-linear equations are known to be often unstable, in the sense that a small variation in the coefficients of the equation (particularly in the coefficient of the highest derivatives) may make a large change to the solution, and secondly, to solve either the weak-field or the full equations requires boundary conditions on the behaviour of the solutions at 'infinity', and there are no rigorously proved results that enable one to make a correspondence between the boundary conditions that seem appropriate in the two theories, the exact and the weak-field theory.

It is therefore best to regard the weak-field solutions not as approximate solutions to the full equations, but as solutions to a different theory that might give an idea of the behaviour expected in the full theory.

4.1.1 Assumptions and approximations

Suppose that there exists some coordinate system in which

$$g_{ij} = \overset{0}{g}_{ij} + h_{ij}$$

where the components h_{ij}, and sufficiently many of their derivatives, are small: say, $|h_{ij}| < \varepsilon$. where ε is some small parameter, and the same for the derivatives of h_{ij}.

We now proceed to approximate Einstein's equations by neglecting terms that are $O(\varepsilon^2)$; approximate equality to this order of accuracy will be denoted by $\approx$.

Firstly,

$$g^{ij} \approx \overset{0}{g}{}^{ij} - h^{ij} \qquad \text{where } h^{ij} := \overset{0}{g}{}^{ik}\,\overset{0}{g}{}^{jl} h_{kl} \tag{1}$$

as can be verified by noting that $(\overset{0}{g}{}^{ij} - h^{ij}) g_{jk} \approx \delta^i_k$. Consequently, if $P_{ij\ldots}$ is a tensor whose components are $O(\varepsilon)$, we can raise an index by writing

$$P^i{}_{j\ldots} = g^{ik} P_{kj\ldots} \approx \overset{0}{g}{}^{ik} P_{kj\ldots}$$

from (1). In other words, raising the index of a small quantity with g has approximately the same effect as raising it with $\overset{0}{g}$. For definiteness, let us now use the convention that indices are always to be raised and lowered with $\overset{0}{g}$.

Inserting these approximations into the definition of the Christoffel symbols (§3.3.10) gives

$$\Gamma^m_{ij} = \left\{ {m \atop ij} \right\} \approx \tfrac{1}{2} \overset{0}{g}{}^{mk} (-h_{ij,k} + h_{jk,i} + h_{ki,j}) \tag{2}$$

so that the definition of Riem given by equation (41) of chapter 3 then gives

$$R^i{}_{ljk} \approx \tfrac{1}{2} \overset{0}{g}{}^{im} (-h_{lk,mj} + h_{km,lj} + h_{lj,mk} - h_{jm,lk}) \tag{3}$$

Contracting by putting $i \to j$ thus gives

$$R_{lk} \approx \tfrac{1}{2}(-h_{lk,}{}^j{}_j + h_k{}^j{}_{,lj} + h_{lj,}{}^j{}_k - h_j{}^j{}_{,lk})$$

We abbreviate this by setting

$$\psi_k := h_k{}^j{}_{,j} - \tfrac{1}{2} h_j{}^j{}_{,k} \tag{4}$$

leading to

$$R_{lk} \approx \tfrac{1}{2}(-h_{lk,}{}^j{}_j + \psi_{k,l} + \psi_{l,k})$$

Finally, inserting this in the field equations (59) of chapter 3 gives

$$8\pi G T_{lk} \approx \tfrac{1}{2}(-h_{lk,}{}^j{}_j + \tfrac{1}{2} h^i{}_{i,}{}^j{}_j \overset{0}{g}_{lk} + \psi_{k,l} + \psi_{l,k} - \psi_{i,}{}^i \overset{0}{g}_{lk}) \tag{5}$$

4.1.2 Gauge transformations

Equation (5) comprises terms in ψ and terms of the form $()_{,}{}^j{}_j$, i.e. of the form $\Delta()$, where Δ is the operator $-\partial^2/\partial x^{0\,2} + \partial^2/\partial x^{1\,2} + \partial^2/\partial x^{2\,2} + \partial^2/\partial x^{3\,2}$ that appears in the wave equation (cf. equation (27) of chapter 2). We now show that we can perform a coordinate transformation so as to remove the ψ terms, leaving a simple wave equation.

The assumptions made at the start of §4.1.1 about the coordinates we are using clearly still leave a lot of freedom. In fact we can make any coordinate transformation we wish, provided that $|\partial x'^i/\partial x^j - \delta^i_j| < \kappa\varepsilon$, together with the derivatives of this equation. So in particular we could use

$$x'^i := x^i + \xi^i \tag{6}$$

where ξ^i and its derivatives are smaller than some multiple of ε. This gives for the transformation matrices

$$L^i{}_j = \frac{\partial x'^i}{\partial x^j} = \delta^i_j + \xi^i{}_{,j} \quad \text{and} \quad (L^{-1})^i{}_j \approx \delta^i_j - \xi^i{}_{,j}$$

Under this transformation the components of g change to

$$\begin{aligned} g'_{ij} &\approx g_{ij} - \xi^k{}_{,i} g_{kj} - \xi^k{}_{,j} g_{ik} \\ &= \overset{0}{g}_{ij} + h'_{ij} \end{aligned}$$

where

$$h'_{ij} := h_{ij} - \xi_{i,j} - \xi_{j,i} \tag{7}$$

So the effect of (6) is to change the value of h, and hence the values of all the quantities derived from it. This sort of transformation is an example of a *gauge transformation*. (Roughly speaking, a gauge transformation is any transformation between mathematically different representations of the same physical situation.)

When h_{ij} changes to the h'_{ij} of (7), then ψ changes to

$$\psi'_i \approx \psi_i - \xi^j{}_{,ij} - \xi_{i,}{}^j{}_j + \xi^l{}_{,li} = \psi_i - \xi_{i,}{}^j{}_j$$

(from the definition (4)). This will be zero if we can choose ξ so that

$$\Delta\, \xi_i \equiv \xi_{i,}{}^j{}_j = \psi_i \tag{8}$$

This equation can always be solved, being just a wave equation with ψ_i as a source term. There is, however, no guarantee that there is a solution that remains 'small' everywhere, although one could be found that was small in some specified bounded region of space–time. But let us assume anyway that this can be done, so that after the transformation the ψ terms no longer appear in (5). (If (8) did not have a small solution, then a more complicated gauge transformation would have to be used, and we would have to handle a more complicated final set of equations.) Thus (5) finally takes the form

$$\tilde{h}_{ij} = -16\pi G T_{ij} \tag{9}$$

where, omitting the prime from h'_{ij} and ψ'_i,

$$\bar{h}_{ij} := h_{ij} - \tfrac{1}{2}h^k{}_k \overset{0}{g}_{ij} \tag{10}$$

or, equivalently (contracting this equation and substituting for $h^k{}_k$)

$$h_{ij} = \bar{h}_{ij} - \tfrac{1}{2}\overset{0}{g}_{ij}\bar{h}^k{}_k \tag{11}$$

The condition $\psi_i = 0$ thus takes the form

$$\bar{h}_i{}^j{}_{,j} = 0 \tag{12}$$

4.1.3 Particle paths

We can substitute the approximation (2) for Γ^i_{jk} in the geodesic equation (14) of chapter 3 to obtain the equation for the world-lines of free particles

$$\frac{d^2x^m}{ds^2} + \tfrac{1}{2}[-h_{ij,}{}^m + 2h_j{}^m{}_{,i}]\frac{dx^i}{ds}\frac{dx^j}{ds} \approx 0 \tag{13}$$

A major simplification occurs if we now restrict attention to slowly moving particles, with $|dx^\alpha/ds| < \varepsilon$, say, $(\alpha = 1, 2, 3)$. We showed earlier (equation (2) of chapter 2) that

$$\frac{dx^0}{ds} = (1 - v^2)^{-\frac{1}{2}}$$

where v is the velocity, and so in the case of slow motion

$$dx^0/ds = 1 + O(\varepsilon^2)$$
$$d^2x^0/ds^2 = O(\varepsilon)$$

Consequently, most of the terms in the summation in the second term of (13) are of order ε^2, the only surviving ones giving

$$\frac{d^2x^\alpha}{dt^2} \approx \frac{d^2x^\alpha}{ds^2} \approx [h_0{}^\alpha{}_{,0} - \tfrac{1}{2}h_{00,}{}^\alpha] \tag{14}$$

where we have put $t := x^0$, identifying the x^0 coordinate with the ordinary laboratory time (at least approximately).

The right-hand-side of (14) now gives us the gravitational acceleration of the particles in our chosen coordinate system.

4.2 Solutions

The wave equation (9) has the well-known solution

$$\bar{h}_{ij}(t, x^1, x^2, x^3) = -\int \frac{4GT_{ij}(t-|\mathbf{x}-\mathbf{y}|, y^1, y^2, y^3)}{|\mathbf{x}-\mathbf{y}|} \mathrm{d}^3\mathbf{y} \tag{15}$$

where

$$|\mathbf{x}-\mathbf{y}| := \left(\sum_{\alpha=1}^{3} (x^\alpha - y^\alpha)^2\right)^{\frac{1}{2}}$$

This represents the field $\bar{h}_{ij}$ as arising from contributions emitted from the source T_{ij} at all points in space, the contribution from each point travelling outward with speed 1 (the speed of light in our units), so that to reach a point x^α at time t it must be emitted from a point y at an earlier time $t-|\mathbf{x}-\mathbf{y}|$.

We now interpret this for a special case.

4.2.1 The Newtonian limit

As in §3.5.2 we now consider the case, appropriate to normal states of matter, in which $\rho = T_{00}$ is the dominant term in T_{ij}, neglecting the role of the other terms. The form (15) now shows that $\bar{h}_{00}$ is the dominant term in $\bar{h}_{ij}$, and so we shall neglect the other terms for now.

In this case we obtain from (11) the relations

$$h_{00} = \bar{h}_{00} + \tfrac{1}{2}h^0{}_0 = \tfrac{1}{2}\bar{h}_{00}$$

$$h_{0\alpha} = 0, \qquad h_{\alpha\beta} = \tfrac{1}{2}\bar{h}_{00}\delta_{\alpha\beta}$$

where, as usual, Greek indices range over the values 1, 2, 3 only, and $\delta_{\alpha\beta}$ denotes the components of the 3×3 unit matrix.

We can obtain an easy interpretation of the field $\bar{h}_{00}$, on noting that the geodesic equation (14) now takes the form

$$\frac{\mathrm{d}^2 x^\alpha}{\mathrm{d}t^2} \approx -\tfrac{1}{4}\frac{\partial}{\partial x^\alpha}\bar{h}_{00}$$

which is identical with the Newtonian expression for the acceleration of a particle in a gravitational field as $-\nabla\phi$ if we interpret $\phi = \bar{h}_{00}/4$ as a gravitational potential.

Returning to the representation (15) of the solution, if we assume that T_{ij} changes very little between times t and $t-|x-y|$ (consistent with the assumption that the matter is moving so slowly that the momentum terms $T_{0\alpha}$ in T_{ij} can be ignored) then we obtain

$$\phi(t, x^\alpha) = \tfrac{1}{4}\bar{h}_{00}(x^i) \approx -G\int \frac{\rho(t, y^\alpha)}{|x-y|}\mathrm{d}^3 y$$

This is again a familiar Newtonian expression: each part of the matter makes a contribution to the gravitational potential that falls off inversely with distance, thus confirming that our choice of field equation (59) of chapter 3 was appropriate.

4.2.2 Gravitational waves

If the source T_{ij} varies periodically with time, then the solution (15) has the form of an outgoing wave. We shall not examine here the way in which the source generates these waves: instead we consider a simple plane wave of the form

$$\tilde{h}_{ij} = \mathscr{R}e\, \tilde{a}_{ij} \exp[ik_l x^l] = \tilde{c}_{ij} \cos(k_l x^l) + \tilde{d}_{ij} \sin(k_l x^l)$$

(where $k_l k^l = 0$), which certainly satisfies the wave equation (9) (cf. §2.3.3). In a small region of space and time any wave-like disturbance will have approximately this form; moreover, by Fourier analysis any solution of (9) can be decomposed into waves of this form, so that there is no loss of generality.

From now on, we shall consider only the cosine term of this wave, the treatment for the sine part being exactly the same.

The gauge condition (12) then demands that

$$\tilde{c}_{ij} k^j = 0 \tag{16}$$

and of course, because h is a perturbation of a symmetric tensor g_{ij}

$$\tilde{c}_{ij} = \tilde{c}_{ji} \tag{17}$$

We now calculate the effect of this gravitational wave on the separation of nearby particles in free fall by using the equation of geodesic deviation for the vector Y that specifies this separation (equation (54) of chapter 3). The geodesic equation (14) will have a solution for which the tangent vector to the path is $X^i = \delta^i_0 + \mathrm{O}(\varepsilon)$ (i.e. the velocity of the particle, as seen in these coordinates, is small) and in this case the geodesic deviation equation becomes

$$\left(\frac{\mathrm{D}^2 Y}{\mathrm{D}s^2}\right)^\alpha \approx R^\alpha{}_{00\beta} Y^\beta \tag{18}$$

The components of Y in a freely falling frame $\{\underset{i}{\bar{E}}\}$ based on the geodesic used for the above equation are given by

$$Y = \bar{Y}^k \underset{k}{\bar{E}}$$

so that

$$g(Y, \underset{i}{\bar{E}}) = \bar{Y}^k \overset{0}{g}_{ik} \Rightarrow \bar{Y}^k = \overset{0}{g}{}^{ik} g(Y, \underset{i}{\bar{E}})$$

(cf. §3.3.1). Differentiating with respect to s twice, and using the fact that $\mathrm{D}\underset{i}{\bar{E}}/\mathrm{D}s = 0$ on the geodesic, $\mathrm{D}g/\mathrm{D}s = 0$ from equation (36) of chapter 3, and $\overset{0}{g}_{ij}$ is constant, gives

$$\frac{\mathrm{d}^2 \bar{Y}^k}{\mathrm{d}s^2} = \overset{0}{g}{}^{ik} g\left(\frac{\mathrm{D}^2 Y}{\mathrm{D}s^2}, \underset{i}{\bar{E}}\right)$$

which is just the expression for a component of $\mathrm{D}^2 Y/\mathrm{D}s^2$ in the freely falling frame. Thus the equation for Y^k in a freely falling frame is obtained simply by transforming (18) into this frame. But the transformation to a freely falling frame only changes a quantity by a fraction of order ε, and hence, to first order in ε, (18) still gives the correct result for a freely falling frame, i.e.

$$\frac{\mathrm{d}^2 \bar{Y}^\alpha}{\mathrm{d}s^2} \approx R^\alpha{}_{00\beta} \bar{Y}^\beta \tag{19}$$

From (11) we see that the metric perturbation h_{ij} is given by

$$h_{ij} = c_{ij} \cos(k_l x^l)$$

where

$$c_{ij} = \tilde{c}_{ij} - \tfrac{1}{2}\tilde{c}^k{}_k \overset{0}{g}_{ij} \tag{20}$$

Thus the expression (3) for the Riemann tensor becomes

$$R^i{}_{ljk} \approx \tfrac{1}{2}\overset{0}{g}{}^{im}(c_{lk}k_m k_j - c_{km}k_l k_j - c_{lj}k_m k_k + c_{jm}k_l k_k)\cos(k_n x^n)$$

from which the equation (19) for Y takes the form

$$\frac{\mathrm{d}^2 \bar{Y}_\alpha}{\mathrm{d}s^2} \approx \tfrac{1}{2}\bar{Y}^\beta(c_{0\beta}k_\alpha k_0 - c_{\beta\alpha}k_0 k_0 - c_{00}k_\alpha k_\beta + c_{0\alpha}k_0 k_\beta) \tag{21}$$

We now simplify this by rotating our coordinates so that the x^1-axis lies in the direction of propagation of the wave, i.e. so that

$$(k_i) = \omega(1, 1, 0, 0) \tag{22}$$

(where ω can be interpreted as the angular frequency of the wave). Then the gauge condition (16) implies that

$$c_{i0} = c_{i1} \tag{23}$$

which in turn gives

$$\begin{aligned} c_{00} &= c_{01} = c_{10} \qquad \text{by symmetry, (17)} \\ &= c_{11} \end{aligned}$$

So (20) becomes

$$c_{ij} = \tilde{c}_{ij} - \tfrac{1}{2}\tilde{c}^{A}{}_{A}\overset{0}{g}_{ij}$$

where from now on capital letters range over the indices 2, 3. Putting this, together with (22) and (23), in (21) then immediately gives

$$\frac{\mathrm{d}^2\bar{Y}_1}{\mathrm{d}s^2} \approx 0, \quad \frac{\mathrm{d}^2\bar{Y}_A}{\mathrm{d}s^2} \approx -\tfrac{1}{2}\bar{Y}^B c_{AB}\cos\omega(x^0+x^1) \tag{24}$$

where

$$c_{AB} = \tilde{c}_{AB} - \tfrac{1}{2}\tilde{c}^{C}{}_{C}\overset{0}{g}_{AB}$$

i.e.

$$c^{A}{}_{A} = 0 \tag{25}$$

To understand what effect this actually has on nearby particles, let us write $[c_{AB}]$ as

$$[c_{AB}] = \begin{bmatrix} \alpha & \beta \\ \beta & -\alpha \end{bmatrix}$$

(which is possible because of (25) and the symmetry (17)). If the initial separation of two particles is $\delta\bar{x}^i = \bar{Y}^i_0$, then the wave will 'wobble' the separation by a variable amount $\delta Y^i := \bar{Y}^i - \bar{Y}^i_0$. If only the α terms are present (with $\beta = 0$) then δY satisfies, from (24),

$$\frac{\mathrm{d}^2\delta Y^2}{\mathrm{d}s^2} = \alpha\omega^2 Y^2_0\cos\omega(x^0+x^1) \tag{26}$$

$$\frac{\mathrm{d}^2\delta Y^3}{\mathrm{d}s^2} = -\alpha\omega^2 Y^3_0\cos\omega(x^0+x^1)$$

while the β terms alone would give

$$\frac{\mathrm{d}^2\delta Y^2}{\mathrm{d}s^2} = \beta\omega^2 Y^3_0\cos\omega(x^0+x^1) \tag{27}$$

$$\frac{\mathrm{d}^2\delta Y^3}{\mathrm{d}s^2} = \beta\omega^2 Y^2_0\cos\omega(x^0+x^1)$$

a general motion being a combination of these two.

The motion (26) is an alternating compression and expansion along the x^2-axis, combined with similar motion with the opposite phase along the x^1-axis. Its effect on a circle of particles in the (x^2, x^3)-plane (the plane transverse to the direction of propagation of the wave) is to deform it as shown in Figure 9. Similarly the effect of (27) is to produce

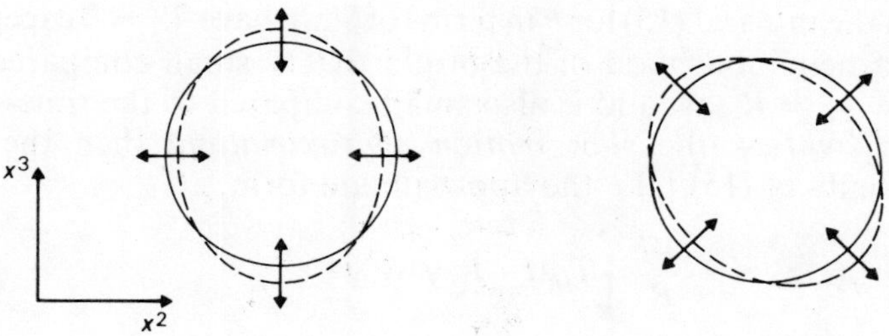

Figure 9

the same pattern of deformation, but along axes that are rotated through 45°. These two motions are thought of as two polarizations of the gravitational wave, analogous to the two polarizations of an ordinary light wave. It is a pattern of motion that is characteristic of any wave that is described by a second-rank tensor.

Many experiments are in progress to detect these waves. Some, currently only at a planning stage, involve the use of pairs of bodies placed in free fall in space, but most use a single elastic body rather than a pair of 'particles'. If we imagine two weights, connected by a spring, being subjected to a wave of the form just discussed, we see that the weights would try to move like free particles, oscillating with the wave, but will be partly restrained by the spring. If the frequency of the wave matches the natural frequency of the mechanical oscillation of the system, then a resonance will build up a large amplitude motion that could be detectable.

In practice a single elastic body (a cylinder of metal or crystal) is used rather than weights and a spring, and the body is based on Earth rather than in free fall, but the basic principles remain the same.

Work of this kind was pioneered by J. Weber at Maryland, who has reported the detection of waves, but other experiments have failed to confirm this, and it appears that more sensitive instruments are required if a certain result is to be achieved.

Exercises

1 Using the approximation of §4.1 (neglecting terms of order ε^2 and assuming $T^i{}_j$ to be of order ε) show from $T_{ij;}{}^j = 0$ that

$$T_{00,}{}^{00} \approx T^{\alpha\beta}{}_{,\alpha\beta} \qquad \text{(i)}$$

Prove also that

$$T^{\alpha\beta}{}_{,\alpha\beta}x^\gamma x^\delta + (T^{\alpha\gamma}x^\delta + T^{\alpha\delta}x^\gamma - T^{\alpha\beta}{}_{,\beta}x^\gamma x^\delta)_{,\alpha} = 2T^{\delta\gamma} \qquad \text{(ii)}$$

(α, β etc run from 1 to 3).

2 If, in the integral (15) for h in terms of T, we have $T_{ij} = 0$ except when $\mathbf{y}$ is in a neighbourhood of the origin that is small compared to the distance $|\mathbf{x}|$, $= R$ say, and is also small compared to the time-scale on which T_{ij} varies (the *slow motion approximation*) then the spatial components of (15) take the approximate form

$$\bar{h}_{\alpha\beta}(t, \mathbf{x}) \approx -\frac{4G}{R} \int T_{\alpha\beta}(t-R, \mathbf{y})\,\mathrm{d}^3\mathbf{y}$$

In this case, show from (i) and (ii) that

$$\bar{h}_{\alpha\beta}(t, x) \approx -\frac{2G}{R} \left.\frac{\mathrm{d}^2 I_{\alpha\beta}}{\mathrm{d}t^2}\right|_{t-R}$$

where

$$I_{\alpha\beta}(t) := \int T_{00}\,(t, \mathbf{y}) y_\alpha y_\beta\,\mathrm{d}^3\,\mathbf{y}$$

is known as the quadrupole moment of the density T_{00}.
(N.B. $\int K_{\dots\alpha,}{}^{\alpha}\mathrm{d}^3 y = \int \mathrm{div}\,(K\dots)\,\mathrm{d}^3 y = 0$, by the divergence theorem, for any quantity $K\dots$ which vanishes outside some bounded region.)

3 A binary star is represented by two equal particles of mass m with world-lines

$$\{(t,\ 0,\ r\cos\omega t,\ r\sin\omega t) : t \in \mathbb{R}^1\}$$

and

$$\{(t,\ 0, -r\cos\omega t, -r\sin\omega t) : t \in \mathbb{R}^1\}$$

Regarding the particles as very small regions of volume V in which $T^{00} = \rho$, with $V\rho = m$, evaluate the quantity $I_{\alpha\beta}$ of Exercise 2, and hence show that, at large distances R from the star

$$[\bar{h}_{AB}] \approx \frac{8mGr^2\omega^2}{R} \begin{bmatrix} \cos 2\omega t & \sin 2\omega t \\ \sin 2\omega t & -\cos 2\omega t \end{bmatrix}$$

$(A, B = 2, 3)$. How does this gravitational wave disturb the relative positions of nearby particles (i) near the z-axis and (ii) in the (x, y)-plane?

4 Write down the weak-field metric for the solution $\bar{h}_{00}(t, \mathbf{x}) = -4Gm/r$ ($r := |\mathbf{x}|$) corresponding to a point mass m at the origin.

A null geodesic in this metric has the form

$$x^i(s) = x_0^i(s) + \xi^i(s)$$

where ξ^i is a small perturbation and x_0^i is the Minkowski space geodesic $x_0^i(s) := s(\delta_0^i + \delta_1^i) + y\delta_2^i$ for some constant y. Ignoring terms in squares of small quantities (i.e. $h_{ij}\xi^k, \xi^l\xi^m$) show that the geodesic equations (13) give in this case

$$\frac{d^2\xi^m}{ds^2} = -\frac{2mGy}{r^3}$$

By considering the direction of the tangent vector $dx_0^i/ds + d\xi^i/ds$ for $s \to \infty$ and $s \to -\infty$, deduce that a light ray passing a distance y from the origin at its closest approach will be deflected by an angle of approximately $4mG/y$ towards the origin.
(This agrees with the exact equations: see § 5.2.2.)

5
Exact solutions

5.1 The Schwarzschild solution

We now turn again to the full field equations (59) of chapter 3 and examine a solution that is static and spherically symmetric, and so corresponds to the gravitational field of a symmetrical non-rotating body, such as a star, with no other bodies nearby.

5.1.1 Symmetry

An exact definition of spherical symmetry would take us beyond the scope of this book. Intuitively, however, we should expect 'spherically symmetric and static' to mean that we could introduce coordinates t, r, θ and ϕ so that the coefficients of the metric were independent of θ, ϕ and t. Actually, this expectation must be modified somewhat: the metric of flat Minkowski space, when transformed into these spherical polar coordinates, takes the form* (see Exercise 1 of chapter 3)

$$\overset{0}{g} = -\mathrm{d}t^2 + \mathrm{d}r^2 + r^2(\mathrm{d}\theta^2 + \sin^2\theta\,\mathrm{d}\phi^2)$$

in which the part $r^2(\mathrm{d}\theta^2 + \sin\theta\,\mathrm{d}\phi^2)$ is just the natural metric on the surface of a sphere of radius r. So in the case of general spherical symmetry we should expect at least this much θ-dependence to be present. Let us assume that dependence on θ and ϕ only enters via this sphere-metric, and apart from this there is no dependence on θ, ϕ or t. Then we are looking for a metric of the form (putting $x^2 = \theta, x^3 = \phi$)

$$g = -v(r)\mathrm{d}t^2 + 2f(r)\mathrm{d}t\mathrm{d}r + p(r)\mathrm{d}r^2 + k_A(r)\mathrm{d}t\mathrm{d}x^A$$
$$+ m_A(r)\mathrm{d}r\mathrm{d}x^A + q(r)(\mathrm{d}\theta^2 + \sin^2\,\theta\,\mathrm{d}\theta^2)\,(A = 2, 3)$$

We now make two further assumptions consistent with the idea of static spherical symmetry.

Firstly, a static solution should be unchanged if we replace t by $-t$. The terms involving f and k_A change sign if this is done, and so must be zero.

Secondly, it must be impossible to select any preferred directions on the spheres $r =$ constant, $t =$ constant. The covector $m_A\mathrm{d}x^A$ does

*Throughout this chapter we use the notation of § 3.3.14 for g.

define such a preferred direction, unless it too is zero. This leaves us with

$$g = -v(r)\mathrm{d}t^2 + p(r)\mathrm{d}r^2 + q(r)(\mathrm{d}\theta^2 + \sin^2 \mathrm{d}\phi^2)$$

where, to ensure a correct identification of time and space, v, p and q must be positive.

A final simplification can be achieved by a coordinate transformation. If we put

$$r' := (q(r))^{\frac{1}{2}}$$

leaving the other coordinates unchanged, and write

$$v(r) = \mathrm{e}^{\nu(r')}, \qquad p(r) = \mathrm{e}^{\lambda(r')}$$

then the final form for g (dropping the prime from r) is

$$g = -\mathrm{e}^{\nu(r)}\mathrm{d}t^2 + \mathrm{e}^{\lambda(r)}\mathrm{d}r^2 + r^2(\mathrm{d}\theta^2 + \sin^2\theta\mathrm{d}\phi^2) = g_{ij}\mathrm{d}x^i\mathrm{d}x^j \qquad (1)$$

5.1.2. Field equations

We now apply the formalism of chapter 3 with the coordinates $x^0 = t$, $x^1 = r$, $x^2 = \theta$, $x^3 = \phi$.

The matrix $\mathbf{g} = [g_{ij}]$ is diagonal (consisting simply of the coefficients in (1)) and so its inverse is found by taking the reciprocals of the diagonal terms, giving

$$g^{00} = -\mathrm{e}^{-\nu}, \quad g^{11} = \mathrm{e}^{-\lambda}, \quad g^{22} = r^{-2}, \quad g^{33} = r^{-2}\sin^{-2}\theta \quad (2)$$

Using the coefficients g_{ij} from (1) in the equations of § 3.3.10 for the Christoffel symbols gives

$$[00,1] = -[01,0] = -\tfrac{1}{2}\mathrm{e}^{\nu}\nu'$$
$$[11,1] = \tfrac{1}{2}\mathrm{e}^{\lambda}\lambda'$$
$$[22,1] = -[21,2] = -r$$
$$[33,1] = -[13,3] = -r\sin^2\theta$$
$$[33,2] = -[32,3] = -r^2\sin\theta\cos\theta$$

and hence from (2)

$$\left\{ {1 \atop 00} \right\} = -\tfrac{1}{2}\mathrm{e}^{\nu-\lambda}\nu' \quad \left\{ {0 \atop 10} \right\} = \left\{ {0 \atop 01} \right\} = \tfrac{1}{2}\nu'$$
$$\left\{ {1 \atop 11} \right\} = \tfrac{1}{2}\lambda' \qquad \left\{ {1 \atop 22} \right\} = -\mathrm{e}^{-\lambda}r$$
$$\left\{ {2 \atop 21} \right\} = \left\{ {2 \atop 12} \right\} = r^{-1} \qquad \left\{ {1 \atop 33} \right\} = -r\mathrm{e}^{-\lambda}\sin^2\theta \qquad (3)$$
$$\left\{ {3 \atop 31} \right\} = \left\{ {3 \atop 13} \right\} = r^{-1} \qquad \left\{ {2 \atop 33} \right\} = -\sin\theta\cos\theta$$
$$\left\{ {3 \atop 32} \right\} = \left\{ {3 \atop 32} \right\} = \cot\theta$$

with the other terms zero.

These expressions are now substituted into the definition (equation (41) of chapter 3) of the Riemann tensor. For example, the component $R^1{}_{212}$ is calculated as

$$\begin{aligned} R^1{}_{212} &= \Gamma^1_{22,1} - \Gamma^1_{21,2} + \Gamma^1_{i1}\Gamma^i_{22} - \Gamma^1_{i2}\Gamma^i_{21} \\ &= \Gamma^1_{22,1} + \Gamma^1_{11}\Gamma^1_{22} - \Gamma^1_{22}\Gamma^2_{21} \quad \text{(all other terms being zero)} \\ &= \tfrac{1}{2}r\lambda' \mathrm{e}^{-\lambda} \end{aligned}$$

where a prime stands for $_{,1} = \partial/\partial r$).

The other components can be calculated similarly, and then contraction gives the components of the Ricci tensor $R_{ij} := R^k{}_{ikj}$ as follows

$$\begin{aligned} R_{00} &= \mathrm{e}^{\nu-\lambda}\left(\tfrac{\nu''}{2} - \tfrac{1}{4}\lambda'\nu' + \tfrac{1}{4}\nu'^2 + \tfrac{\nu'}{r}\right) \\ R_{11} &= -\tfrac{\nu''}{2} + \tfrac{1}{4}\nu'\lambda' - \tfrac{1}{4}\nu'^2 + \tfrac{\lambda'}{r} \\ R_{22} &= \mathrm{e}^{-\lambda}\left(\tfrac{1}{2}\tfrac{\lambda'}{r} - \tfrac{1}{2}\tfrac{\nu'}{r} - 1\right) - 1 \\ R_{33} &= R_{22}\sin^2\theta \end{aligned} \tag{4}$$

with the rest zero. From these the Ricci scalar is

$$R := R^i{}_i = -\mathrm{e}^{-\lambda}\left(\nu'' - \tfrac{1}{2}\lambda'\nu' + \tfrac{1}{2}\nu'^2 + \tfrac{2\nu'}{r} - \tfrac{2\lambda'}{r} + \tfrac{2}{r^2}\right) + \tfrac{2}{r^2} \tag{5}$$

We now apply the field equations (Einstein's equations) (59) of chapter 3. At this stage it is useful to compare the energy momentum tensor T with the special relativity case, and so give a physical interpretation of its components. To do this, as we discussed in §3.3.1, we must transform to a freely falling frame $\bar{\mathbf{x}}$. The only restriction on choosing such a frame at one point is the condition that the metric have components $\overset{0}{g}_{ij}$, i.e. that the frame vectors $\underset{i}{\bar{E}}$ satisfy $g(\underset{i}{\bar{E}}, \underset{j}{\bar{E}}) = \overset{0}{g}_{ij}$, and this is seen to be satisfied by

$$\underset{0}{\bar{E}}{}^i = \mathrm{e}^{-\nu/2}\delta^i_0, \quad \underset{1}{\bar{E}}{}^i = \mathrm{e}^{-\lambda/2}\delta^i_1, \quad \underset{2}{\bar{E}}{}^i = r^{-1}, \quad \underset{3}{\bar{E}}{}^i = r^{-1}(\sin\theta)^{-1}\delta^i_3$$

The components of T in this frame are the numbers $T(\underset{i}{\bar{E}}, \underset{j}{\bar{E}})$ (see §3.3.1), which give ρ, p_1, p_2, p_3 (the density and pressures: see §2.2.4) since in our case the tensor is diagonal. Thus we have

$$\rho = T(\underset{0}{\bar{E}}, \underset{0}{\bar{E}}) = \mathrm{e}^{-\nu}T_{00} = -T^0{}_0$$

and similarly $p_1 = T^1{}_1, \quad p_2 = T^2{}_2, \quad p_3 = T^3{}_3$.

Putting this, together with (4) and (5), in Einstein's equation (59) of chapter 3 gives

$$-8\pi G\rho = -r^{-2}+r^{-2}e^{-\lambda}(1-r\lambda') \tag{6a}$$

$$8\pi G p_1 = -r^{-2}+r^{-2}e^{-\lambda}(1+r\nu') \tag{6b}$$

$$8\pi G p_2 = 8\pi G p_3 = e^{-\lambda}(-\tfrac{1}{2}\nu''-\tfrac{1}{4}\nu'^2-\tfrac{1}{2}\nu' r^{-1}+\tfrac{1}{2}\lambda' r^{-1}+\tfrac{1}{4}\lambda'\nu') \tag{6c}$$

Equation (6a) can now be solved for λ in terms of ρ. For, writing the equation as

$$-8\pi G\rho = r^{-2}\frac{d}{dr}(e^{-\lambda}r-r)$$

gives

$$r(e^{-\lambda}-1) = -2m(r)G \tag{7}$$

where

$$m(r) := \int_0^r 4\pi r'^2 \rho \, dr'$$

(assuming that $e^{-\lambda}$ is bounded at $r = 0$, as it must be if the geometry is to be well defined there). The function m introduced here is an expression that, in special relativity, would be the total energy (mainly in the form of mass) inside radius r. But this interpretation is slightly misleading in general relativity: $4\pi r^2 dr$ is not the volume of a shell of radius r and thickness dr because the form of the metric implies that a coordinate displacement δr has an actual length of $e^{\lambda/2}\,\delta r$.

Using (7) in the metric (1) then gives

$$g_{11} = e^{\lambda} = 1/(1-2m(r)G/r) \tag{8}$$

If we knew the pressures p_α we could continue and evaluate ν, but this would require a detailed knowledge of the nature of the matter making up the body. However, if we are dealing with a body such as a star in which all the matter is concentrated inside the sphere $r = R$, then we can find ν outside the star by setting $T = 0$ there. For $r > R$ we have

$$m(r) = \int_0^R 4\pi r'^2 \rho dr' = m(R) = : M \qquad \text{say,}$$

and so (6b) with $p_\alpha = 0$ gives

$$\left(1-\frac{2MG}{r}\right)(1+r\nu') = 1$$

which integrates to

$$e^{-\nu} = K/(1-2MG/r)$$

K is a constant of integration that we must set to one if we are to obtain a metric that tends to the Minkowski form at large distances. Indeed K can be absorbed by rescaling the coordinate t.

Finally, putting this with (8) in (1) gives

$$g = -(1-2MG/r)\mathrm{d}t^2 + (1-2MG/r)^{-1}\mathrm{d}r^2 + r^2(\mathrm{d}\theta^2 + \sin^2\theta\mathrm{d}\phi^2) \qquad (9)$$

valid outside the body. This metric is known as the Schwarzschild metric.

5.1.3 Eddington's extension of the metric

The expression (9) is obviously only defined for $r \neq 2MG$: as r decreases to this critical radius the coefficients g_{11} and g_{00} tend to infinity and zero respectively. The *Schwarzschild radius* $r_s = 2MG$ is, for most astronomical bodies, comparatively small. For the Sun $r_s = 2.96$ km and for the Earth $r_s = 0.88$ cm; in both cases r_s is much less than the radius R of the body itself, and inside the body the coefficient g_{11} takes the form (8) which in these cases remains bounded. This will not, however, be the case for very large or very dense bodies having $2MG > R$, and for these we need to investigate the significance of the behaviour at r_s.

We can make a coordinate transformation (described below) so that the metric in the new coordinates is regular at $r = r_s$. The apparent 'bad behaviour' of (9) is due to our having used coordinates that are not appropriate to describing this region. (The situation is analogous to the use of polar coordinates on the sphere, metric $g = \mathrm{d}\theta^2 + \sin^2\theta\,\mathrm{d}\phi^2$, for which $g^{22} = \sin^{-2}\theta$ tends to infinity at the poles, even though there is nothing singular there geometrically.) When continued inside $r = r_s$ the metric in fact no longer remains static (though it is still spherically symmetric) and so it is not surprising that our coordinates, that were chosen on the assumption of a static metric, cannot describe this region adequately.

We introduce a new coordinate u so that the ingoing null geodesics (light rays) travelling radially with constant θ and ϕ lie in the surfaces $u =$ constant (see Figure 10(a)). For a radial null geodesic with tangent vector components $X^i = \mathrm{d}x^i/\mathrm{d}s$ we have

$$\mathbf{X} = \begin{pmatrix} \mathrm{d}x^0/\mathrm{d}s \\ \mathrm{d}x^1/\mathrm{d}s \\ 0 \\ 0 \end{pmatrix}$$

and $0 = g(X,X) = -(1-2MG/r)\,(\mathrm{d}x^0/\mathrm{d}s)^2 + (1-2MG/r)^{-1}\,(\mathrm{d}x^1/\mathrm{d}s)^2$

i.e.

$$\frac{\mathrm{d}x^1}{\mathrm{d}x^0}\left(=\frac{\mathrm{d}r}{\mathrm{d}t}\right) = \pm\left(1-\frac{2MG}{r}\right)$$

Choosing the negative sign (so that r decreases with t, corresponding to ingoing null geodesics) gives

$$t = -(r + 2MG\ln|r - 2MG|) + k$$

where k is a constant of integration). Thus the geodesics lie in surfaces of constant u, where

$$u = t + r + 2MG\ln|r - 2MG| \tag{10}$$

Consequently

$$dt = -dr(1 - 2MG/r)^{-1} + du$$

so that the metric (9) takes the form

$$g = 2dudr - (1 - 2MG/r)du^2 + r^2(d\theta^2 + \sin^2\theta d\phi^2) \tag{11}$$

Although the coefficient of du^2 still becomes zero at r_s we have removed the unboundedness there. The form (11) is now acceptable because it can easily be shown (see Exercise 2) that at every point in the range $0 < r < \infty$ a frame can be chosen in which $g'_{ij} = \overset{0}{g}_{ij}$.

This extension of the space–time beyond r_s by using different coordinates is due to Eddington.

We have included the modulus signs in the coordinate transformation (10) so that it is equally applicable to (9) in the region where $0 < r < r_s = 2MG$. So in fact the expression (9) describes the whole of the metric (11) for all r, except on the hypersurface $r = r_s$. The

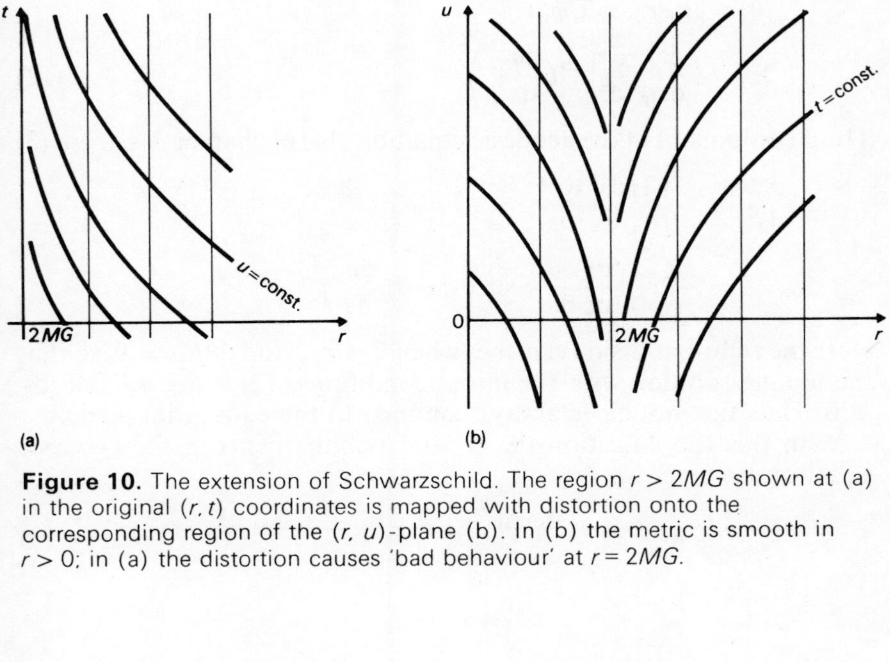

Figure 10. The extension of Schwarzschild. The region $r > 2MG$ shown at (a) in the original (r, t) coordinates is mapped with distortion onto the corresponding region of the (r, u)-plane (b). In (b) the metric is smooth in $r > 0$; in (a) the distortion causes 'bad behaviour' at $r = 2MG$.

relationship between the two coordinate systems is shown in Figure 10.

We notice that in the inner region $0 < r < r_s$ the coefficients of dt^2 and dr^2 have the 'wrong' signs: t is here a spacelike coordinate and r becomes a timelike coordinate. In this region the space–time, while still entirely in accord with our basic mathematical principles, can no longer be called static since the time coordinate is r and g_{ij} depends on r. A space–time which, well away from a spherical source, is static will thus stop being static at any points closer to the source than r_s (but still outside the matter); in this region the space–time evolves with decreasing r until after a finite time (i.e. a finite interval of r) the metric coefficients tend to infinity at $r = 0$. This time we reach a genuine singularity that cannot be removed by another coordinate transformation, as shown by the scalar $R_{ijkl}R^{ijkl}$, which does not depend on coordinates and tends to infinity.

5.1.4 Schwarzschild geodesics

We now solve the geodesic equations (14) of chapter 3 to determine the paths of light rays and particles in this metric, particles traversing timelike geodesics and light rays traversing null ones (§3.3.11)

We can at once simplify the calculation by noting that, if $X_0 = \dot{\gamma}(0)$ is the tangent vector at the start of a geodesic (the point where a particle is released or a light pulse emitted), then we can use the spherical symmetry to perform a spatial rotation of the coordinates to ensure that the spatial components of X are tangential to the equatorial plane $\theta = \pi/2$, i.e. so that

$$\gamma(0) = (t_0, r_0, \pi/2, \phi_0)$$

$$\mathbf{X}_0 = \left(\left.\frac{dt}{ds}\right|_0, \left.\frac{dr}{ds}\right|_0, 0, \left.\frac{d\phi}{ds}\right|_0\right)^{T} \tag{12}$$

The θ-component of the geodesic equation (14) of chapter 3 is, from (3)

$$\frac{d^2\theta}{ds^2} = -\begin{Bmatrix} 2 \\ ij \end{Bmatrix}\frac{dx^i}{ds}\frac{dx^j}{ds}$$

$$= -2r^{-1}\frac{dr}{ds}\frac{d\theta}{ds} + \sin\theta\cos\theta\left(\frac{d\phi}{ds}\right)^2$$

Here the right-hand-side vanishes when $\theta = \pi/2$ and $d\theta/ds = 0$, so that the unique solution with the initial conditions (12) is $\theta \equiv \pi/2$; $d\theta/ds \equiv 0$. Thus the motion is always confined to the equatorial plane.

With this simplification the ϕ- and t-components of the geodesic equation become

$$\frac{d^2\phi}{ds^2} = -2r^{-1}\frac{d\phi}{ds}\frac{dr}{ds} \tag{13}$$

or

$$\frac{d}{ds}\ln\frac{d\phi}{ds} = -2\frac{d}{ds}\ln r$$

and

$$\frac{d^2t}{ds^2} = -\begin{Bmatrix}0\\ ij\end{Bmatrix}\frac{dx^i}{ds}\frac{dx^j}{ds} = -\nu'\frac{dt}{ds}\frac{dr}{ds} \tag{14}$$

or

$$\frac{d}{ds}\left(\ln\frac{dt}{ds}\right) = -\frac{d\nu}{ds}$$

Thus both (13) and (14) can be integrated to give

$$\frac{d\phi}{ds} = \frac{h}{r^2} \qquad (h \text{ constant}) \tag{15}$$

and

$$\frac{dt}{ds} = E\left(1-\frac{2MG}{r}\right)^{-1} \qquad (E \text{ constant}) \tag{16}$$

It is actually no coincidence that these equations can be integrated exactly: we can show (Exercise 6 of chapter 3) that for each symmetry of the metric one obtains an integral of the geodesic equation, and (15) and (16) arise from the symmetries in ϕ and t respectively. We now apply the condition (§3.3.11) for a geodesic to be timelike or null, which is

$$\begin{aligned} g(\dot\gamma.\dot\gamma) &= -\left(1-\frac{2MG}{r}\right)\left(\frac{dt}{ds}\right)^2 + \left(1-\frac{2MG}{r}\right)^{-1}\left(\frac{dr}{ds}\right)^2 + r^2\left(\frac{d\phi}{ds}\right)^2 \\ &= -E^2\left(1-\frac{2MG}{r}\right)^{-1} + \left(\frac{dr}{ds}\right)^2\left(1-\frac{2MG}{r}\right)^{-1} + \frac{h^2}{r^2} \qquad (17) \\ &= k = \begin{cases} -1 & \text{(timelike)} \\ 0 & \text{(null)} \\ 1 & \text{(spacelike)} \end{cases} \end{aligned}$$

giving

$$\left(\frac{dr}{ds}\right)^2 = \left(k-\frac{h^2}{r^2}\right)\left(1-\frac{2MG}{r}\right) + E^2 \tag{18}$$

This equation could now be integrated to find $r(s)$, and we should then be able to integrate (15) and (16) (in principle) to give the complete solution of the problem. In practice, it is enough to extract the main features of the solution.

Before continuing, let us interpret the constants h and E introduced in (15) and (16), for the case of geodesics that come from, or escape to, large distances. For large values of r, (16) gives

$$\frac{dt}{ds} \to E \qquad (r \to \infty) \tag{19}$$

But at large distances the metric tends to the Minkowski form, and so it is reasonable to apply the special relativity interpretation here.

In the case of a timelike geodesic, we find from §2.1.2 that a particle of mass m moving on the geodesic has a total energy of

$$P^0 = \frac{m dt}{ds} \approx mE \qquad \text{(from (19))}$$

so that in this case E is the *energy per unit mass* of a particle traversing the geodesic, as measured at large r.

In the case of a null geodesic, we have from §2.3.3 (cf. the remarks in §3.3.11) that dt/ds, and hence E, can be chosen to give the *frequency* of the light wave whose wave-fronts are defined by the geodesic. It is interesting that in quantum theory the frequency is proportional to the energy of the photons which make up the light in the quantum picture, so here again E is an energy.

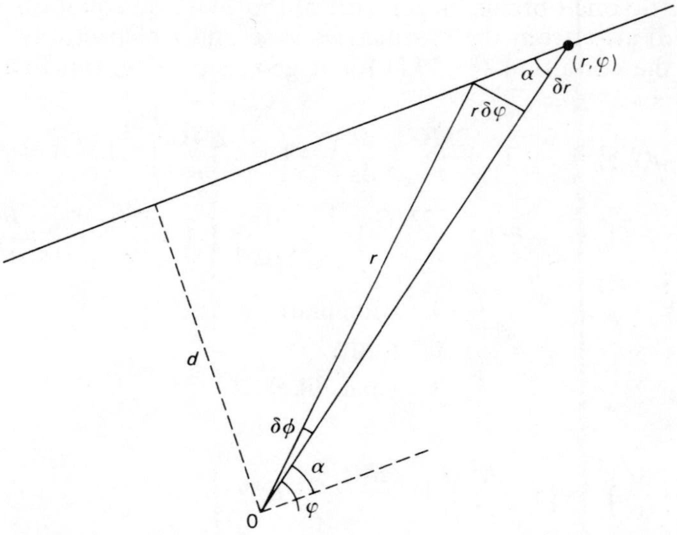

Figure 11. The impact parameter d is the distance from O to the tangent to the trajectory, in the limit as $r \to \infty$. The tangent is determined by δr and $\delta\phi$ ($\delta\phi \to 0$). From the figure, $d = r \sin \alpha$ where $\tan \alpha = r d\phi/dr$; hence $d = r^2 (d\phi/dr) \cos \alpha \to r^2 d\phi/dr$ (as $r \to \infty$ and $\alpha \to 0$).

From (15) and (18) we find that, as $r \to \infty$, for $k = -1$

$$\frac{\mathrm{d}\phi}{\mathrm{d}r} \approx \frac{h}{r^2\sqrt{(E^2-1)}}$$

This specifies a curve in the (r, ϕ)-plane whose tangent passes at a distance d from the origin, where

$$d = \frac{h}{\sqrt{(E^2-1)}}$$

(see Figure 11). The distance d, called the *impact parameter*, can be interpreted as the distance at which the particle would pass the origin if it were moving in ordinary Minkowski space, with no gravitational field. In this case we should assign to a particle of mass m an angular momentum of

$$(m^*v)d \approx m\frac{\mathrm{d}r}{\mathrm{d}s}d \approx m\sqrt{(E^2-1)}\,.\,d = mh$$

m^* is the apparent mass and v is the velocity. See §2.1.2.) Thus h is the angular momentum per unit mass.

For a null geodesic (light ray) it is better to use the impact parameter itself as the most direct interpretation of h. Repeating the above arguments in this case gives

$$h = dE \tag{20}$$

for the relation between h and the impact parameter d.

Bearing this in mind, we now return to examine the solutions of (18). The case $h = 0$ corresponds to $\phi =$ constant, where the geodesic runs straight in to $r = 0$. So we now concentrate on the more interesting case $h \neq 0$. We use a graphical method suited to any equation of the form

$$\left(\frac{\mathrm{d}r}{\mathrm{d}s}\right)^2 = E^2 - f(r) \tag{21}$$

where, in our case

$$f(r) \equiv f(r; h, k) = \left(\frac{h^2}{r^2} - k\right)\left(1 - \frac{2MG}{r}\right)$$

For each E the allowed range of r is that for which the right-hand-side of (21) is positive (since the left is a square), i.e. where $E^2 \geq f(r)$. This is exhibited by drawing the line $y = E^2$ parallel to the x-axis on the same diagram as a graph of $y = f(x)$; the allowed range of r is represented by the part of this line that lies above the graph (shown continuous in Figure 12). The solutions then fall into various types according to the

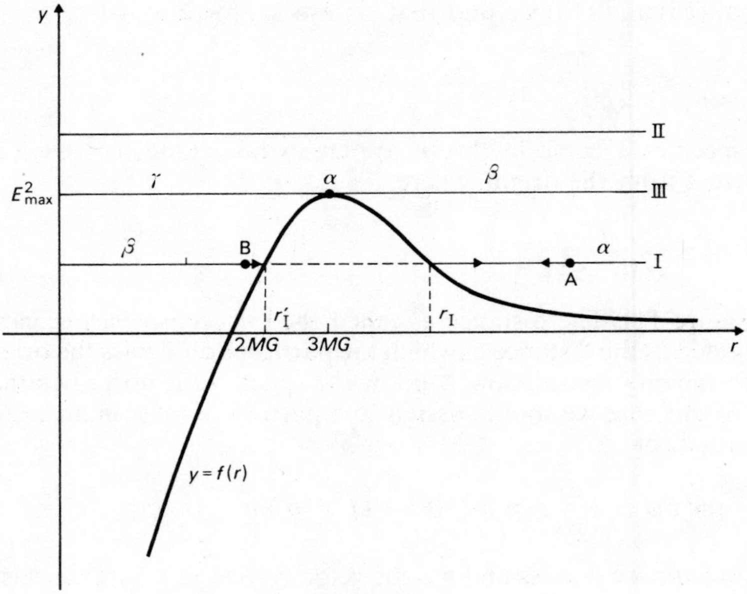

Figure 12

relation of this line to the graph of f. We consider the various cases in turn.

(A) *Null geodesics* ($k = 0$)
The graph of f in this case is shown in Figure 12. It has a maximum of $E^2_{max} = h^2/27M^2G^2$ at $r = 3MG$ and becomes negative at $2MG$, tending to $-\infty$ as $r \to 0$. For large r it asymptotes to zero.

(I) $0 < E^2 < E^2_{max}$
In this case the line $y = E^2$ (labelled I on the figure) has two sections (shown continuous) where $E^2 \geq f$, say $r \geq r_I$ and $0 < r \leq r'_I$. Two types of motion correspond to this:
(α) $r \geq r_1$. Suppose the photon (particle of light) is initially moving inwards—to the left on the figure with $dr/ds < 0$—starting at A, say. The photon moves in until it nears $r = r_I$. At r_I we should have $E^2 = f$ and so (from (21)) $dr/ds = 0$, i.e. the photon becomes stationary. To see what happens we differentiate (21) to give

$$\frac{d^2r}{ds^2} = -\tfrac{1}{2}f'$$

This shows that near r_I the acceleration d^2r/ds^2 is strictly positive, and so the photon slows down its r-motion, stops at r_I (when it is then moving only in the ϕ direction) and then moves outwards again.

Throughout the motion it is moving round in ϕ according to (15), and so the net result is that the photon traverses the path in the (r, ϕ)-plane shown in Figure 13. If the photon is initially moving outwards, it traverses only the latter part of this path.

(β) $0 < r \leq r'_1$. By a similar argument, light travelling initially outwards, starting at B, say, attains a maximum radius r' and then reverses and goes inwards to the singularity at $r = 0$.

(II) $E^2 > E^2_{max}$

The light goes monotonically inwards to $r = 0$ or, with the opposite initial velocity, outwards to ∞.

(III) $E^2 = E^2_{max}$

(α) $r \equiv 3MG$. At $3MG$ we have $d^2r/ds^2 = f' = 0$ and so a solution is possible in which r remains constant at this value: the light moves in a circular orbit at fixed r.

(β) $r > 3MG$. Light moving inwards approaches $3MG$ asymptotically,

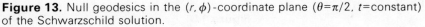

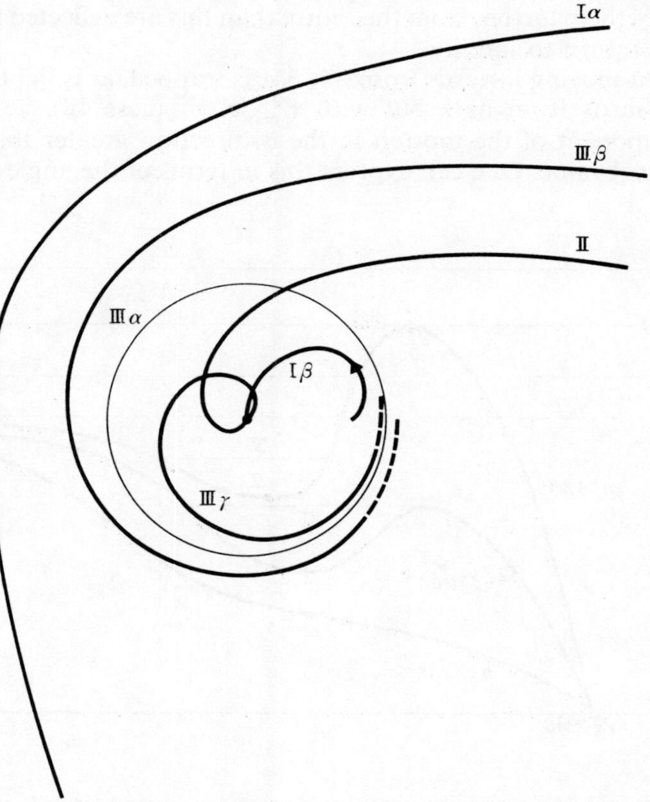

Figure 13. Null geodesics in the (r, ϕ)-coordinate plane ($\theta = \pi/2$, t=constant) of the Schwarzschild solution.

giving the spiral orbit shown in Figure 13, which may be traversed in either direction.
(γ) $r < 3MG$. In this case the circular orbit is approached from the inside.

An important point arises in the cases Iβ, II and IIIγ when the initial point lies inside the region $r < 2MG$. Here r is a time coordinate (§5.1.3) which decreases to the future. If we adopt the convention that s increases to the future (as we have done up to now, so that $\mathrm{d}r/\mathrm{d}s < 0$ means that the geodesic is moving inwards) then we must have $\mathrm{d}r/\mathrm{d}s < 0$ when $r < 2MG$. Geodesics in this region must necessarily proceed to $r = 0$ and cannot get outside the Schwarzschild radius $2MG$.

The main features that emerge from this classification of null geodesics are the following.

(1) Light moving inwards from $r > 3m$ and having $E^2 > E^2_{\mathrm{max}}$ (case II) is trapped. This condition is $E^2 > d^2E^2/27M^2G^2$, i.e. the impact parameter d is less than $3\sqrt{3}MG$. Rays whose initial direction takes them further from the centre than this are deflected (case Iα) and return to infinity.

(2) Light moving inwards from $r \leq 3M$ is trapped, as is light moving outwards from here but with $E^2_{\mathrm{max}} > E^2$ (case Iβ), i.e. with a component of the motion in the ϕ-direction greater than some critical value. One can express this in terms of the angle that the

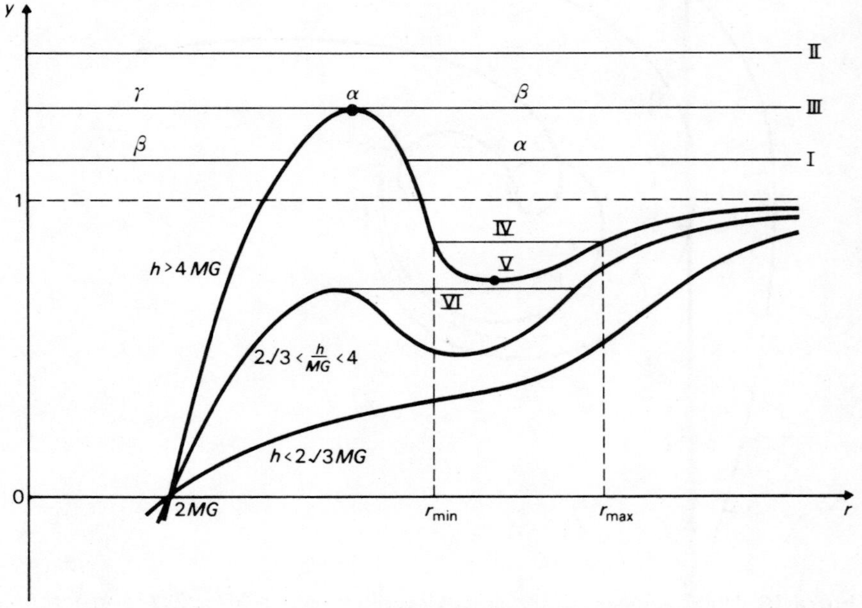

Figure 14

photon must initially make with the outward radius if it is to escape, and it can be shown (Exercise 3) that this angle tends to zero as $r \to 2MG$.

(3) Beyond this point light cannot escape: the geodesic proceeds with the time coordinate r until $r = 0$.

Because of this trapping of light, the Schwarzschild metric with no matter outside $r = 2MG$ is termed a black hole.

(B) *Timelike geodesics* ($k = -1$)
The graphs of f in this case are sketched in Figure 14 for values of h typical of the various cases that arise. The new feature present is a minimum in the cases $h > 2\sqrt{3}MG$. The maximum is also present for just this range of h, and the value at the maximum, E^2_{max}, is greater than the asymptotic value 1 when $h > 4MG$.

For suitable h and E all types of orbit that we found in the null case are possible; examples are indicated on the diagram by corresponding numbers. In addition we have several new types.

(IV) A planetary type of orbit, the particle revolving round the source between two limiting values of r, $r_{min} \leq r \leq r_{max}$.

(V) A limiting case of IV when r_{min} and r_{max} coincide and we have a stable circular orbit at constant r at the minimum of f. (The orbit IIIα is unstable, because a small perturbation could turn it into IIIγ.)

(VI) An orbit for $2\sqrt{3}MG < h < 4MG$ that spirals away from the unstable circular orbit, reaches a maximum r and then spirals down again. (i.e. it combines features of IIIβ and IV.)

The overall picture is the same as for null geodesics: particles become progressively more likely to be trapped as they near $r = 2MG$.

5.2 Experimental tests of general relativity

The more remarkable orbits in Figure 13 would provide a very good test of relativity theory, but they could only be observed for black holes or for bodies that were so compact that they were almost entirely inside their Schwarzschild radius $2MG$. However, the orbits of type Iα and IV can be examined for ordinary bodies such as the Sun, and provide the most important tests of general relativity.

During a total eclipse of the Sun the stars whose rays to the Earth pass close to the Sun become visible. According to § 5.1.4. the light from these stars should be deflected by the Sun, so that the relative positions of the stars that are near the Sun in the sky should be distorted: the apparent angular position of a star should be displaced outwards by the angle by which the light is deflected. This can be calculated for null geodesic orbits of type I.

A second test can be applied, though less directly, to the planetary orbits (type IV) in the solar system. Unlike the elliptical Newtonian orbits these do not return to their starting point after ϕ has increased by 2π. So on each revolution the point at which r is a minimum (the *perihelion*) occurs at a slightly different value of ϕ. The change in this value per revolution (a quantity called the *precession* of the perihelion) can be calculated and compared with observation.

Both these effects can best be analysed in terms of the perturbation which general relativity makes to the Newtonian orbits. Finally, in §5.2.4, we shall examine a third test involving the frequency, rather than the path, of light.

5.2.1 Perturbation of Newtonian orbits

Putting $u := 1/r$ in equation (18) to give

$$\left(\frac{\mathrm{d}u}{\mathrm{d}s}\right)^2 = u^4\left(\frac{\mathrm{d}r}{\mathrm{d}s}\right)^2 = u^4\{(k-h^2u^2)(1-2MGu)+E^2\}$$

whence, from (15)

$$\left(\frac{\mathrm{d}u}{\mathrm{d}\phi}\right)^2 = \left(\frac{\mathrm{d}u}{\mathrm{d}s}\right)^2 \Big/ h^2u^4 = (E^2+k)/h^2 - (2MGk/h^2)u - u^2 + 2MGu^3 \qquad (22)$$

We now show that, in the solar system, the last term can be neglected in comparison with the u- and u^2-terms. Comparing the last term with u^2 we have

$$\frac{2MGu^3}{u^2} = \frac{2MG}{r} = \frac{r_s}{r} \lesssim 10^{-5}$$

The u term enters only for timelike geodesics, when its relation to the last term is

$$\frac{2MGu^3h^2}{2MGu} = \frac{h^2}{r^2} = r^2\left(\frac{\mathrm{d}\phi}{\mathrm{d}s}\right)^2 \lesssim \frac{v^2}{c^2} \lesssim 10^{-5}$$

We therefore expect a solution of the form $u = u_0 + u_1$, where u_1 is a small perturbation and u_0 is a solution of (22) with the last term omitted, i.e.

$$\left(\frac{\mathrm{d}u_0}{\mathrm{d}\phi}\right)^2 = (E^2+k)/h^2 - (2MGk/h^2)u_0 - {u_0}^2 \qquad (23)$$

This is in fact precisely the Newtonian equation. If we ignore squares of small quantities we have the approximation

$$\left(\frac{\mathrm{d}u}{\mathrm{d}\phi}\right)^2 \approx \left(\frac{\mathrm{d}u_0}{\mathrm{d}\phi}\right)^2 + 2\left(\frac{\mathrm{d}u_0}{\mathrm{d}\phi}\right)\left(\frac{\mathrm{d}u_1}{\mathrm{d}\phi}\right)$$

and inserting this and (23) in (22) gives

$$\frac{\mathrm{d}u_0}{\mathrm{d}\phi}\frac{\mathrm{d}u_1}{\mathrm{d}\phi} \approx -\frac{MGk}{h^2}u_1 - u_0u_1 + MGu_0{}^3 \tag{24}$$

where terms of the form $2MGu_0{}^2u_1$ etc. have been ignored as being a product of the two small quantities $2MGu_0{}^3$ and u_1/u_0.

5.2.2 The deflection of light

We now insert $k = 0$ into the equations of the previous section. The Newtonian equation (23) becomes

$$\frac{\mathrm{d}u_0}{\mathrm{d}\phi} = (A^2 - u_0{}^2)^{\frac{1}{2}} \qquad (A := E/h = 1/d)$$

where we can, without loss of generality, take the positive root. (The negative root simply corresponds to the light going round in the opposite direction.) Integrating gives

$$u_0 = (1/d)\sin(\phi - \phi_0) \tag{25}$$

or $\qquad r\sin(\phi - \phi_0) = d$

Note that this is simply a straight line at an angle of ϕ_0 to the x-axis and passing the origin at a distance d. We can rotate the origin of ϕ to ensure that $\phi_0 = 0$ in what follows.

Putting (25) and its derivative $\mathrm{d}u_0/\mathrm{d}\phi = (1/d)\cos\phi$ in (24) gives

$$\cos\phi\frac{\mathrm{d}u_1}{\mathrm{d}\phi} \approx -\sin\phi \,.\, u_1 + \frac{MG}{d^2}\sin^3\phi$$

which can be written

$$\cos^2\phi\frac{\mathrm{d}}{\mathrm{d}\phi}\frac{u_1}{\cos\phi} \approx \frac{MG}{d^2}\sin^3\phi$$

integrating to

$$u_1 \approx \frac{MG}{d^2}[1 + \cos^2\phi + a\cos\phi]$$

where a is a constant of integration.

Adding u_0 and inverting gives

$$r = d(\sin\phi + (MG/d)[1 + \cos^2\phi + a\cos\phi])^{-1} \tag{26}$$

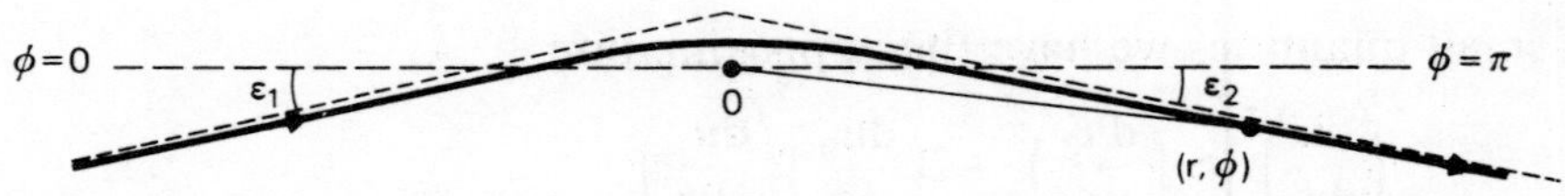

Figure 15. The deflection of light. ε_1 and ε_2 are the angles between the asymptotic directions and the line $\phi=0/\phi=\pi$. O is the point $r=0$ and a radius vector is shown from O to a typical point (r, ϕ) on the geodesic. ϕ tends to $-\varepsilon_1$ $(r \to -\infty)$ and $\pi+\varepsilon_2$ $(r \to \infty)$, and the deflection is $\varepsilon_1+\varepsilon_2$.

To calculate the deflection in the orbit we require the asymptotic values of ϕ for large values of r(see Figure 15). Clearly $r \to \infty$ as the part in parentheses in (26) tends to zero, i.e. the asymptotic values of ϕ are the roots of

$$\sin\phi + (MG/d)[1+\cos^2\phi + a\cos\phi] = 0 \tag{27}$$

Since MG/d is small, these roots are almost 0 and π: we write them as $+\varepsilon_1$ and $\pi-\varepsilon_2$. Substituting in (27) and making the appropriate 'small angle' approximations gives

$$\varepsilon_1 + (MG/d)[2+a] = 0$$
$$\varepsilon_2 + (MG/d)[2-a] = 0$$

The total deflection $\delta\phi$ is, from Figure 15, $\varepsilon_1+\varepsilon_2$, and adding these equations gives

$$\delta\phi = \varepsilon_1+\varepsilon_2 \approx 4MG/d$$

This is about 10^{-5} rad for a ray of light just grazing the Sun.

The result has been confirmed by observations of the stars during eclipses to an accuracy of around 10%, and by observations of the apparent positions of radio sources to about one per cent.

5.2.3 The precession of the perihelion

For timelike geodesics ($k = -1$) the Newtonian equation (23) gives

$$\frac{du_0}{d\phi} = (C^2 - v^2)^{\frac{1}{2}}$$

where $v := u_0 - MG/h^2$ and $C^2 := (E^2-1)/h^2 + M^2G^2/h^4$

Integration gives

$$v = C\sin(\phi-\phi_0) \tag{28}$$

or

$$\frac{1}{C} - r\sin(\phi-\phi_0) = \left(\frac{MG}{h^2C}\right)r$$

Note that the left-hand-side is the distance of the point (r, ϕ) from a straight line l distant $1/C$ from the origin making an angle ϕ_0 with the x-axis. And so the equation represents a conic section with l as directrix and the origin as focus; the eccentricity is

$$\varepsilon := \frac{Ch^2}{MG} \tag{29}$$

For the bounded orbits of type IV we require $E^2 < 1$ (see Figure 14) and so $C < MG/h^2$ and $\varepsilon < 1$: we have an ellipse as the Newtonian approximation.

We now rotate the ϕ-coordinate as before so as to make $\phi_0 = 0$. From (28) and (29) u_0 is given by

$$u_0 = \frac{MG}{h^2}(1 + \varepsilon \sin \phi)$$

so

$$\frac{du_0}{d\phi} = \frac{MG\varepsilon}{h^2} \cos \phi$$

Inserting these expressions into (24) gives, after cancellation,

$$\cos \phi \frac{du_1}{d\phi} \approx - \sin \phi u_1 + \frac{(MG)^3}{h^4 \varepsilon}(1 + \varepsilon \sin \phi)^3$$

This can be integrated in the same way as the equation for u_1 in the previous subsection, to give

$$u_1 \approx a \cos \phi + \frac{(MG)^3}{h^4}[(1/\varepsilon + 3\varepsilon) \sin \phi + 3 + \varepsilon^2 + \varepsilon^2 \cos \phi \\ - 3\varepsilon\phi \cos \phi]$$

The term $a \cos \phi$ can be absorbed into the $\varepsilon \sin \phi$ of u_0 by redefining ε and ϕ while most of the remaining terms are insignificantly small, being multiplied by the factor $(MG)^3/h^4$ which is completely negligible in comparison with u_0. The exception is the $\phi \cos \phi$ term, which increases steadily as ϕ increases and, though multiplied by the same very small factor, can eventually become arbitrarily large. It is this that is responsible for the steady precession of the perihelion of the orbit.

Retaining only this term gives a total solution

$$u \approx \frac{MG}{h^2}(1 + \varepsilon \sin \phi - 3(MG/h)^2 \varepsilon\phi \cos \phi)$$

$$\approx \frac{MG}{h^2}(1 + \varepsilon \sin \{\phi - 3(MG/h)^2 \phi\})$$

We see from this that u goes from one maximum value (perihelion) to the next when the term $\phi(1 - 3(MG)^2/h^2)$ increases by 2π. Thus the perihelion advances by an approximate amount $6\pi(MG/h)$ per revolution.

To detect this we need a planet which is near the Sun, so that the gravitational effects are large (i.e. h is relatively small), and which has a fairly large eccentricity of orbit so that the position of perihelion can accurately be ascertained. Mercury satisfies both these criteria; moreover, it had been known for some time before the advent of general relativity that its perihelion did precess by a greater amount than could be explained by the perturbations of other planets. The value of this unexplained precession was in remarkable agreement with the general relativistic prediction.

While this is welcome confirmation, it should not be assumed that the matter is quite as straightforward as this account might suggest. The agreement comes from combining the traditional Newtonian analysis with a general relativistic correction: ideally a consistently general relativistic calculation should be done, but this is impossible at present. Moreover, there are many other effects (such as a slight departure of the Sun from a perfect sphere) that could produce a similar result.

5.2.4 The gravitational red shift

We now show how the equation for null geodesics (light rays) leads to an important experimental test of general relativity.

As remarked in § 5.1.4 a comparison of general relativity in a freely falling frame $\bar{\mathbf{x}}$ with the special relativity form for an electromagnetic wave (§ 2.3.3) indicates that, for a suitable choice of the affine parameter on the geodesic (which is only determined up to a linear transformation, § 3.3.11) we may take $\bar{X}^0$, the time component of the tangent vector in a freely falling frame, to be the frequency of the wave in whose wave-front the geodesic lies. Thus the observed frequency of a light ray depends both on the point on the geodesic at which the tangent X is evaluated and the choice of a freely falling observer at that point.

Suppose, for the sake of definiteness, that the source of the field is on the Earth, and that we are considering the transmission of light from one point to another. We shall see that it is particularly interesting to have the two points at different distances r from the centre. We are thus considering observers at constant r, and if the effect of the rotation of the Earth be momentarily neglected, constant θ and ϕ. Such observers are not themselves freely falling, but we can use the frames of freely falling observers that instantaneously coincide with those at constant r when they receive or send out the light. So we need to use frames at the two events of emitting and receiving the light for which $\underset{0}{\bar{E}}$ is tangent to the constant-r-θ-ϕ world-lines. Such a frame is the one used after (5) to evaluate T, for which $\underset{0}{\bar{E}}{}^i = e^{-\nu/2}\delta^i_0$. Indeed, all we need to do is multiply the tangent vector to the world-line with r, θ and ϕ constant, with components δ^i_0, by a factor such that it has a length-squared of -1.

So, if X is the tangent vector to the null geodesic describing the light ray, the frequency in this frame is

$$n = \bar{X}^0 = \overset{00}{g}{}^{0i} g(X, \underset{i}{\bar{E}}) \qquad \text{(cf. § 3.3.1)}$$
$$= e^{\nu/2} X^0 = E(1-2MG/r)^{-\frac{1}{2}} \qquad \text{(from (16))}$$

Thus the ratio of the frequencies at two points on the ray with $r = r_1$ and $r = r_2$ is

$$\frac{n_1}{n_2} = \sqrt{\left(\frac{1-2MG/r_2}{1-2MG/r_1}\right)}$$

We have already noted that, for the Earth, MG is about 0.5 cm, so that MG/r is very small (about 7×10^{-10}). Thus we can write approximately

$$\frac{\delta n}{n_2} \approx \frac{MG}{r_1}\cdot\frac{\delta r}{r_2}$$

where $\delta n := n_1 - n_2$ and $\delta r := r_2 - r_1$. If we replace MG/r^2 by the Newtonian value of the acceleration due to gravity at the surface of the Earth, g, and use laboratory units in which the velocity of light c is no longer 1, then the change in frequency becomes

$$\frac{\delta n}{n} \approx \frac{g\delta r}{c^2} \tag{30}$$

This effect was detected in the laboratory by using as a source of electromagnetic radiation a nuclear transition that produced radiation at a particularly pure frequency. It was transmitted upwards several metres and compared with the frequency associated with the same transition in the new position. Good agreement with the above formula was obtained.

It should be pointed out, however, that this experiment is actually not independent of the Eötvös experiment because (30) can easily be deduced from the principle of equivalence. All we need do is to argue that the laboratory on the Earth is equivalent to a laboratory in free space that is accelerating upwards with acceleration g. In such a case, when the light is received the laboratory will be moving faster than when it was emitted, by an amount $g\delta r/c$, and the resulting relative velocity between the source when the light is emitted and the receiver when it is received gives a Doppler shift of $g\delta r/c^2$.

5.3 Exact gravitational waves

In § 4.2.2 we examined waves based on functions of the form

$$f = A \cos(k_i x^i) = A \cos \omega u \tag{31}$$

(where A is constant, $u := x^0 + x^1$, k is a constant covector and we rotate and boost the coordinates so that $\mathbf{k} = (1, 1, 0, 0)$). This satisfies the wave equation

$$g^{ij} f_{,ij} \equiv \left\{ -\left(\frac{\partial}{\partial x^0}\right)^2 + \left(\frac{\partial}{\partial x^1}\right)^2 + \left(\frac{\partial}{\partial x^2}\right)^2 + \left(\frac{\partial}{\partial x^3}\right)^2 \right\} f = 0 \tag{32}$$

We can see that any function $g(u)$ also satisfies this equation, since

$$\frac{\partial^2}{\partial x^{0\,2}} g(u) = \frac{\partial^2}{\partial x^{1\,2}} g(u) = g''(u)$$

(indeed, any function can be Fourier analysed into a combination of cos u and sin u terms). Thus (31) can be regarded as the simplest form of a plane wave with a sinusoidal shape, while $g(u)$ represents a more general plane wave with an arbitrary shape.

We can go further and allow the amplitude of the wave to vary at different points on its wave-front, while keeping the same overall shape, by considering a function of the form*

$$h(x) = A(x^2, x^3) g(u) \tag{33}$$

which will satisfy (32) provided that

$$\left(\frac{\partial}{\partial x^2}\right)^2 A + \left(\frac{\partial}{\partial x^3}\right)^2 A = 0 \tag{34}$$

The form (33) is called a *plane-fronted wave*. It represents a disturbance propagating with fixed shape and amplitude in the negative-x^1 direction, the wave-fronts (the surfaces u = constant at any time x^0) being planes.

We now look for a corresponding solution in general relativity.

5.3.1 Assumptions for plane-fronted waves

To begin with, put $v := x^0 - x^1$, $u := x^0 + x^1$. Then we carry over the following features of (33).

(i) The wave-fronts u = constant, v = constant are planes perpendicular to the lines of u and v with the Euclidean metric $(dx^2)^2 + (dx^3)^2$. This forces the metric to have the form $h du^2 + p dv^2 + 2k du dv + (dx^2)^2 + (dx^3)^2$.

*In fact we can use $h(x) \equiv h(x^2, x^3, u)$ subject to (34).

(ii) The tangent vector to the lines of constants u, x^2 and x^3 (i.e. the direction of propagation) is a null vector. To consider the implications of this, let us relabel our coordinates, putting now

$$x^0 \equiv u, \qquad x^1 \equiv v \tag{35}$$

Then the direction of propagation is the vector X with components $X^i = \delta^i_1$ and length $g(X, X) = g_{11} = p$. So X will be null provided that $p = 0$.

(iii) The components are independent of v.

If the remaining components h and k depended only on u (so that we had a plane wave) then we could remove k by defining a new coordinate u' so that $\mathrm{d}u' = k\mathrm{d}u$. In the present case we cannot do this without also introducing extra $\mathrm{d}x^A\mathrm{d}u$-terms. Let us, however, simplify the problem by making the arbitrary assumption that $k = 1$ (which will be justified when we later find that we can have a solution of this form that is precisely analogous to (33)). Thus we finally consider the metric

$$g = h\mathrm{d}u^2 + 2\mathrm{d}u\mathrm{d}v + (\mathrm{d}x^2)^2 + (\mathrm{d}x^3)^2 \tag{36}$$

5.3.2 Application of field equations

We now look for a pure gravitational wave, propagating in a vacuum. Putting $T = 0$ in equation (61) of chapter 5 gives Ric $= 0$, which will lead to an equation for h.

With the labelling (35) of coordinates, the metric coefficients in (36) are

$$g_{00} = h, \quad g_{01} = 1 = g_{22} = g_{33} \tag{37}$$

with the rest zero. So the components of the inverse metric are

$$g^{11} = -h, \quad g^{10} = g^{22} = g^{33} = 1 \tag{38}$$

with the rest zero. Using (37) and (38) in the formulae for the Christoffel symbols and the Riemann tensor (equations (29) and (41) of chapter 3) then quickly leads to

$$R^2{}_{020} = R^1{}_{202} = -\tfrac{1}{2}h_{,22}$$

$$R^3{}_{030} = R^1{}_{303} = -\tfrac{1}{2}h_{,33}$$

$$R^1{}_{230} = R^1{}_{320} = R^2{}_{003} = R^3{}_{002} = \tfrac{1}{2}h_{,23}$$

with $R^i{}_{jkl} = -R^i{}_{jlk}$ and all other components zero.

The vacuum condition is then

$$0 = R_{jl} := R^i{}_{jil} = -\tfrac{1}{2}(h_{,22} + h_{,33})\delta^0_j\delta^0_l$$

which corresponds to (34). Thus to each plane-fronted wave of the form (33), there is a corresponding vacuum gravitational wave. There exist plane-fronted gravitational waves with arbitrary profiles $g(u)$ and any amplitude $A(x^2, x^3)$ that is a solution of (34), for which the metric (36) has

$$h = A(x^2, x^3)g(u)$$

Note, however, that $A = $ constant is simply flat space in a distorted coordinate system, since in this case Riem $= 0$.

5.4 Cosmological models

5.4.1 Basic assumptions

A cosmological model is one that represents the overall structure of the entire universe. As a first step (and the only step that so far is solidly established) we aim to model only the very large-scale features.

We find from observational astronomy that, on a sufficiently large scale, the universe is *isotropic*: it appears much the same in all directions with no tendency for galaxies or distant radio sources to be concentrated in any one part of the sky. We represent this in terms of spherical symmetry, as with the Schwarzschild solution. But it is unlikely that we should be precisely at the centre of a universe that is merely spherically symmetric: a more reasonable assumption is that the universe is *homogeneous*, i.e. every point in space is equivalent and on the whole the universe looks the same to all observers and is spherically symmetric about any one of them.

The idea of homogeneity assumes that there is some way of separating space out of space–time. More precisely, we assume that there is a time-coordinate t such that on each spacelike surface $S_\tau := \{x : t(x) = \tau\}$ all points are equivalent. At every point on this surface there could be an observer, such as ourselves, whose frame $\{\underset{i}{E}\}$ has $\underset{0}{E}$ perpendicular to S (so that this surface really is his 'space') and to whom the universe appears isotropic. We can imagine the universe filled with the world-lines of these hypothetical *fundamental observers*, or, more realistically, we can identify these fundamental observers with the galaxies. Our model universe will be mathematically homogeneous and isotropic, although these features are present in the real universe only in the sense of averages over very large regions.

We now look for a form of metric that embodies these assumptions. It can be proved rigorously that the form chosen is the only one in which the surfaces S_τ are homogeneous and have isotropy about every point, though we shall not attempt to do this.

Choosing some particular time τ_0 and putting coordinates x^1, x^2, x^3 on S_{τ_0}, we can then extend these coordinates to the other S_τ by projecting

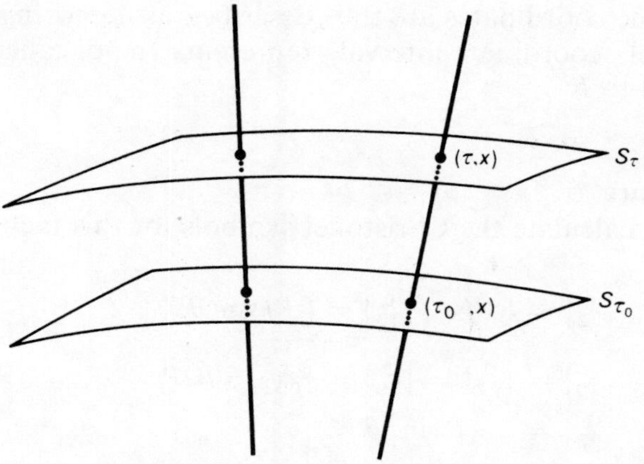

Figure 16. Two hypersurfaces (with one dimension omitted) with constant time coordinates in a homogeneous cosmology. The world-lines of fundamental observers link points with the same spatial coordinates.

along the world-lines of the fundamental observers, requiring that they have constant values of these spatial coordinates. (Such coordinates are called *comoving*: see Figure 16.) If one observer is selected as origin, we may choose the coordinates to have a spherical polar form about this observer; spherical symmetry and the condition that the world lines of fundamental observers (with r, θ and ϕ constant) be perpendicular to the S_τ gives the metric in the form

$$g = -V\mathrm{d}t^2 + p(r,t)\mathrm{d}r^2 + q(r,t)(\mathrm{d}\theta^2 + \sin^2\theta\mathrm{d}\phi^2)$$

We now use the idea of homogeneity. Firstly, this implies that V depends only on t, and so we can redefine t so as to absorb this factor (i.e. set $\mathrm{d}t' = V\mathrm{d}t$ for a new coordinate t'). So we assume V to be unity. Secondly, the equivalence of all observers implies that if at some time observers A and B are the same distance apart as observers P and Q then this will remain the case. This implies that the only possible change in time in the spatial part of the metric is an overall change in scale by a time-dependent factor* that we designate $(R(t))^2$. Finally, we redefine the r-coordinate so that at τ_0 the coefficient of $\mathrm{d}r^2$ is 1. This gives

$$g = -\mathrm{d}t^2 + (R(t))^2[\mathrm{d}r^2 + f(r)(\mathrm{d}\theta^2 + \sin^2\theta\mathrm{d}\phi^2)] \tag{39}$$

This metric describes a family of observers which move with the matter in the universe, and, while keeping their relative distances in the same proportions, suffer a progressive *expansion* (or contraction) governed by the scale-factor R. The observers carry their coordinates

*This factor R (a terminology that is unfortunately fairly standard) is *not* the Ricci scalar $R^i{}_i$.

with them (the coordinates are thus described as comoving) so that a given spatial coordinate-interval represents a distance that is proportional to R.

5.4.2 Curvature

As usual, we calculate the Christoffel symbols for this metric, finding

$$\left\{ {0 \atop 11} \right\} = R\dot{R} \qquad \left\{ {0 \atop 22} \right\} = R\dot{R}f \qquad \left\{ {0 \atop 33} \right\} = R\dot{R}f\sin^2\theta$$

$$\begin{aligned}
&\left\{ {1 \atop 10} \right\} = \left\{ {1 \atop 01} \right\} = \left\{ {2 \atop 20} \right\} = \left\{ {2 \atop 02} \right\} = \left\{ {3 \atop 30} \right\} = \left\{ {3 \atop 03} \right\} = \dot{R}/R \\
&\left\{ {1 \atop 22} \right\} = -\tfrac{1}{2}f' \quad \left\{ {1 \atop 33} \right\} = -\tfrac{1}{2}f'\sin^2\theta \\
&\left\{ {2 \atop 12} \right\} = \left\{ {2 \atop 21} \right\} = \left\{ {3 \atop 13} \right\} = \left\{ {3 \atop 31} \right\} = \tfrac{1}{2}f'/f \\
&\left\{ {2 \atop 33} \right\} = -\sin\theta\cos\theta \quad \left\{ {3 \atop 23} \right\} = \cot\theta
\end{aligned} \tag{40}$$

with the rest zero.

From these we can calculate the Ricci tensor, which turns out to be diagonal, with components

$$\begin{aligned}
R^0{}_0 &= \frac{3\ddot{R}}{R} \\
R^1{}_1 &= \frac{2\dot{R}^2}{R^2} + \frac{\ddot{R}}{R} + \frac{1}{R^2}\left(\frac{f'^2}{2f^2} - \frac{f''}{f}\right) \\
R^2{}_2 &= \frac{2\dot{R}^2}{R^2} + \frac{\ddot{R}}{R} + \frac{1}{R^2}\left(\frac{1}{f} - \frac{f''}{2f}\right) \\
R^3{}_3 &= R^2{}_2
\end{aligned} \tag{41}$$

We can repeat the argument of § 5.1.2 to relate this to the freely falling frame $\{\underset{i}{\bar{E}}\}$ for which

$$\underset{0}{\bar{E}}{}^i = \delta^i_0, \quad \underset{1}{\bar{E}}{}^i = R^{-1}\delta^i_1, \quad \underset{2}{\bar{E}}{}^i = R^{-1}f^{-\frac{1}{2}}\delta^i_2, \quad \underset{3}{\bar{E}}{}^i = R^{-1}f^{-\frac{1}{2}}(\sin\theta)^{-1}\delta^i_3$$

As in the discussion following (5) we find that the components $\bar{R}_{ij} = R(\underset{i}{\bar{E}}, \underset{j}{\bar{E}})$ in this frame (which can be regarded as the frame of a fundamental observer) are equal to the type (1, 1) components listed in (41) above.

The condition of isotropy thus demands that the spatial components are equal: $R^1{}_1 = R^2{}_2 = R^3{}_3$. Homogeneity requires that they be independent of r, so we must have

$$f'^2/2f^2 - f''/f = \kappa \tag{42a}$$

$$1/f - f''/2f = \kappa \tag{42b}$$

with κ a constant.

We now solve these equations, subject to the condition that $\{r = 0\}$ corresponds to a point in space, the origin of coordinates of a spherical polar type, which implies that $f(0) = 0$.

We consider two cases, with $\kappa = 0$ and $\kappa \neq 0$.

(i) $\kappa = 0$
Equation (42b) gives

$$f = r^2 + Cr + D$$

and the condition on $f(0)$ gives $D = 0$. Substituting in (42a) then gives $C = 0$, i.e. $f = r^2$. The spatial part of the metric (apart from the scale factor R^2) is thus

$$\mathrm{d}r^2 + r^2(\mathrm{d}\theta^2 + \sin^2\theta \mathrm{d}\phi^2)$$

which is the metric of Euclidean 3-space in spherical polar coordinates. In this case space is flat, though space–time is curved.
(ii) $\kappa \neq 0$
Here the solution to (42b) is

$$f(r) = 1/\kappa + A\mathrm{e}^{r\sqrt{-2\kappa}} + B\mathrm{e}^{-r\sqrt{-2\kappa}} \tag{43}$$

Once again, substituting in (42a) gives

$$AB = 1/4\kappa^2$$

Putting, therefore, $B = 1/4A\kappa^2$ allows us to write (43) in the form

$$f(r) = A[\mathrm{e}^{r\sqrt{-\kappa/2}} + (1/2A\kappa)\mathrm{e}^{-r\sqrt{-\kappa/2}}]^2$$

Finally, applying the condition $f(0) = 0$ gives $A = -1/2\kappa$ and thus the solutions

$$f(r) = \frac{2}{\kappa}\sin^2\sqrt{\left(\frac{\kappa}{2}\right)}r \quad \text{when} \quad \kappa > 0 \tag{44}$$

and

$$f(r) = -\frac{2}{\kappa}\sinh^2\sqrt{\left(\frac{-\kappa}{2}\right)}r \quad \text{when} \quad \kappa < 0 \tag{45}$$

We still have some freedom left in the choice of coordinates: we can rescale r so as to absorb the square-root factor by setting

$$r^* := \sqrt{(|\kappa|/2)} \,.\, r$$

$$R^* := \sqrt{(2/|\kappa|)} \,.\, R$$

(with the convention that $r^* = r$ and $R^* = R$ in the case $\kappa = 0$). Then the metric takes the form

$$g = -\mathrm{d}t^2 + (R^*(t))^2[\mathrm{d}r^{*2} + s(r^*)(\mathrm{d}\theta^2 + \sin^2\theta\mathrm{d}\phi^2)]$$

where

$$s(r^*) = \begin{cases} r^{*2} & (\kappa = 0) \\ \sin^2 r^* & (\kappa > 0) \\ \sinh^2 r^* & (\kappa < 0) \end{cases}$$

In the case $\kappa \neq 0$, therefore, space as well as space–time is curved. The case $\kappa > 0$ is of particular geometrical interest. Here, as r^* increases from 0, the area of the sphere $r^* =$ constant, $t =$ constant, with metric

$$\sin^2 r^*(\mathrm{d}\theta^2 + \sin^2\theta\mathrm{d}\phi^2)$$

increases to a maximum at $r^* = \pi/2$ and then starts to decrease, becoming zero (i.e. shrinking to a point, as at $r = 0$) when $r^* = \pi$. This behaviour is analogous to the way in which the circles $\theta =$ constant on a sphere (the lines of latitude) become larger as one moves from the North Pole to the equator and then shrink to the South Pole. Indeed, the analogy is exact because the spatial metric

$$\mathrm{d}r^{*2} + \sin^2 r^*(\mathrm{d}\theta^2 + \sin^2\theta\mathrm{d}\phi^2)$$

turns out to be the metric of a three-dimensional sphere (i.e. the hypersurface $x^2 + y^2 + z^2 + w^2 = 1$ in $\mathbb{R}^4$). In this case space closes up on itself at large distances and has finite total volume.

5.4.3 The cosmological red shift

As we did for the Schwarzschild solution in § 5.2.4, we shall now compare the frequencies at two points on a null geodesic, as measured by fundamental observers. If we identify the fundamental observers with the galaxies, this corresponds to a comparison of the frequencies at which light is emitted at one galaxy and received at another, distant, galaxy (such as our own).

Because of homogeneity and isotropy, all null geodesics are equivalent, so it is sufficient to consider a radial null geodesic for which θ and ϕ are constant (conditions that are easily seen to be compatible with the geodesic equations). Then the r-component of the geodesic equation (14) of chapter 3 is, from (40)

$$\frac{\mathrm{d}^2 r}{\mathrm{d}s^2} + \frac{2\dot{R}}{R}\left(\frac{\mathrm{d}r}{\mathrm{d}s}\right)\left(\frac{\mathrm{d}t}{\mathrm{d}s}\right) = 0$$

which integrates (because the homogeneity implies that there is a symmetry which, on the geodesic, corresponds to displacement in r; see the remarks on equations (15) and (16)) giving

$$\frac{\mathrm{d}r}{\mathrm{d}s} = a(R(t))^{-2} \qquad (a \text{ constant}) \tag{46}$$

The condition that the geodesic be null is

$$g(X,X) = -\left(\frac{\mathrm{d}t}{\mathrm{d}s}\right)^2 + (R(t))^2\left(\frac{\mathrm{d}r}{\mathrm{d}s}\right)^2 = 0 \tag{47}$$

which, together with (46) gives the 0-component of the tangent vector as

$$X^0 = \frac{\mathrm{d}t}{\mathrm{d}s} = \frac{a}{R(t)} \tag{48}$$

Now, as we saw in §5.2.4, the frequency of the light ray associated with this geodesic as seen by an observer with a freely falling frame $\{\underset{i}{\bar{E}}\}$ is

$$n = \bar{X}^0 = \overset{0}{g}{}^{0i}\, g(X, \underset{i}{\bar{E}}) = -g(X, \underset{0}{\bar{E}})$$

For a fundamental observer the components of $\underset{0}{\bar{E}}$ in our coordinates are $\underset{0}{\bar{E}}{}^i = \delta^i_0$, and so this equation is simply

$$n = X^0 = \frac{a}{R(t)}$$

from (48). Thus the ratio of the frequencies measured at two points (t_1, r_1, θ, ϕ) and (t_2, r_2, θ, ϕ) is

$$\frac{n_1}{n_2} = \frac{R_2}{R_1} \tag{49}$$

where $R_1 := R(t_1)$ and $R_2 := R(t_2)$.

If the point labelled 1 is the point where the light is emitted, and point 2 where it is received, then we see from this that the light will be received at a lower frequency than that at which it was emitted ($n_2 < n_1$) if the scale factor of the universe R is increasing ($R_2 > R_1$). It is in fact observed that the light from more distant galaxies is reddened, i.e. that the frequency of spectral lines associated with known atomic transitions is observed at a lower frequency than the frequency of emission for that transition. This means, in terms of this model, that R is increasing, or the universe is expanding.

In practice, optical observations are confined to comparatively nearby galaxies, so that we can approximate our formulae by taking the intervals Δr and Δt to be 'small' (i.e. we neglect their squares), where

$$\Delta r := |r_2 - r_1|$$

$$\Delta t := t_2 - t_1$$

The red shift Δz is then usually expressed in terms of the relative frequency change

$$\Delta z := \frac{n_1 - n_2}{n_2} = \frac{R_2}{R_1} - 1 \tag{50}$$

from (49). But for small intervals we have

$$R_2 \approx R_1 + \dot{R}(t_2)\Delta t$$

and from (47) $\Delta t \approx R_1 \Delta r$
so (50) becomes

$$\Delta z \approx \dot{R}\Delta r$$

giving a red shift that increases (linearly because we are neglecting higher order terms) with increasing Δr.

This cannot be checked directly by observation because r is merely a mathematical coordinate. However the quantity $\Delta l := R\Delta r$ represents a length in the spatial part of the metric and can be correlated with observational measures of distance such as the apparent brightness or apparent diameter of the source of light. So the relation between red shift and the 'distance' Δl is

$$\Delta z = H\Delta l \quad \text{with } H := \dot{R}/R \tag{51}$$

where H is a constant named Hubble's constant, after the first observer of this effect.

If we include higher order terms then the relation is no longer linear but depends in more detail on the shape of $R(t)$. Moreover, we should then have to specify what observational measure of distance we were going to use, since different measures would correspond to different functions of Δr and so to different distance/red-shift relations.

5.4.4 Field equations

We now repeat the development that led to equation (6), substituting (41) in the field equations (59) of chapter 3 to find the pressures and density in terms of the metric. The result is

$$8\pi G\rho = -8\pi G T^0{}_0 = \frac{3R^2}{R^2} + \frac{3\kappa}{2R^2} = \frac{3\dot{R}^{*2}}{R^{*2}} + \frac{3k}{R^{*2}} \tag{52}$$

$$8\pi G p = 8\pi G T^1{}_1 = -\frac{2\ddot{R}}{R} - \frac{\dot{R}^2}{R^2} - \frac{\dot{\kappa}}{2R^2} = -\frac{2\ddot{R}^*}{R^*} - \frac{\dot{R}^{*2}}{R^{*2}} - \frac{k}{R^{*2}} \tag{53}$$

where $k = \text{sgn}(\kappa) = 0$, 1 or -1, p is the pressure (the same in all directions) and ρ is the density (i.e. energy-density) of our smoothed-out model, representing the average pressure and density in the universe. A dot represents the derivative $\mathrm{d}/\mathrm{d}t$.

Eliminating k between (52) and (53) gives

$$\frac{8\pi G}{3}(\rho + 3p) = -2\ddot{R}^*/R^*$$

Here the mass density is certainly positive, and it turns out that for all normal types of matter the pressure p is either positive or no more negative than $-\rho/3$. So let us assume that $\rho + 3p > 0$. Then the above equation gives

$$\ddot{R}^* < 0 \tag{54}$$

We have already seen that, because of the evidence of the cosmological red shift, the expansion rate $\dot{R}^*(t_2)$ is positive at the time t_2 in the model representing the present time. So if $\dot{R}^*$ were constant at its present value, R^* would have to become zero at a time

$$t_H := R^*(t_2)/\dot{R}^*(t_2) = R(t_2)/\dot{R}(t_2)$$

ago (where t_H is called the Hubble time, equal to the reciprocal of Hubble's constant; see Figure 17). But the result (54) shows that the graph $R^*(t)$ curves downwards, $\dot{R}^*(t)$ having been larger in the past. Thus R^* will have become zero at a time t_0 between $t_2 - t_H$ and the present time t_2.

As t decreases to t_0 the factor R^*, giving the separation of the fundamental observers, shrinks to zero. The matter in the universe (which is at rest relative to the fundamental observers) is concentrated more and more densely near t_0—indeed we shall show below that the

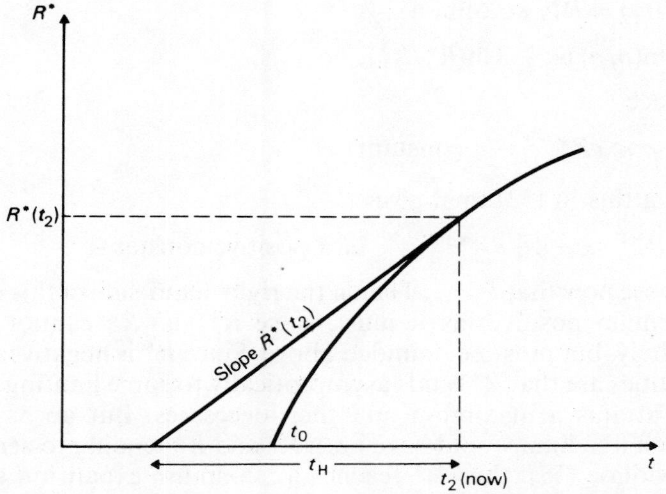

Figure 17. The diagram gives t_H from the relation $\dot{R}^*(t_2) = R^*(t_2)/t_H$. Since $\ddot{R}^* < 0$ (the graph curves down), R^* becomes zero less than a time t_H ago.

density is unbounded there and so we can speak of a real singularity at the start of the universe. This is the 'big bang' model: to describe it in conventional terms, as t increases from t_0 the matter progressively expands, the density falling and the rate of expansion decreasing (equation (54)) throughout.

Given that the expansion of the universe is slowing down, it is possible that the expansion may eventually stop, and perhaps turn into contraction, or it may continue expanding forever. The result depends on the constant k. We shall show that for $k = 1$ (when space is a finite-volume three-dimensional sphere) the expansion reverses and the universe recontracts, while in the other cases (where space is infinite) the expansion continues indefinitely.

Differentiating (52) with respect to t and eliminating $\ddot{R}^*$ and k from (52) and (53) gives

$$\dot{\rho}+3(p+\rho)\dot{R}^*/R^* = 0 \qquad (55)$$

This is actually the t-component of the conservation equation $T^{ij}{}_{;j} = 0$, which, as we have seen in §3.5.3, is a consequence of the field equations.

We now make stronger assumptions about the matter, in order to put bounds on p. Let us assume first that the pressure is positive: a greater restriction than $\rho+3p > 0$, but still one that is satisfied by matter in normal conditions, and which is certainly valid for matter at the present well-expanded state of the universe. In this case (55) gives

$$\dot{\rho} < -3\rho\dot{R}^*/R^*$$

If we integrate this up from a lower limit at a time t_1 when $\rho(t_1) = \rho_1$ and $R^*(t_1) = R_1^*$, we obtain

$$\ln(\rho/\rho_1) < -3\ln(R^*/R_1^*)$$

and hence

$$\rho < cR^{*-3} \quad (c \text{ constant})$$

Inserting this in (52) then gives

$$\dot{R}^{*2} < -k+dR^{*-1} \qquad (d \text{ a positive constant})$$

Suppose now that $k = 1$. Then if the right-hand-side of this equation is to remain positive (as it must, since $\dot{R}^{*2}$ is) R^* cannot increase indefinitely, but must be bounded above. Since $\ddot{R}^*$ is negative, the only possibilities are that R^* tends asymptotically to some limiting value, or that it attains a maximum and then decreases. But an asymptotic approach to a limit would have both $\dot{R}^*$ and $\ddot{R}^*$ tending to zero, which would violate (53); thus R^* reaches a maximum expansion and then decreases (see Figure 18).

To show the converse, we need an inequality in the opposite

direction for p. We shall assume that $p < \rho$. Again, this is true for the low density matter in the present epoch of the universe. Alternatively, it could be deduced from the assumption that the speed of sound in the matter, given by $(\mathrm{d}p/\mathrm{d}\rho)^{\frac{1}{2}}$, is less than the speed of light, which is unity.

Then (55) gives

$$\dot{\rho} > -6\rho\dot{R}^*/R^*$$

and hence, integrating up from time t_1 as before

$$\rho > c'R^{*-6}$$

(This, incidentally, confirms the picture that the density becomes unbounded when $R^* \to 0$.)

Thus (52) becomes

$$\dot{R}^{*2} > -k + d'R^{*-4} \qquad (d' \text{ a positive constant})$$

It follows immediately from this that if k is -1 or zero (an infinite universe), then the factor R^* increases to infinity.

In principle it should be possible to tell from higher terms in the distance/red-shift relation what the function $R^*(t)$ is, and hence whether the universe is finite in space or infinite. But in practice such a direct determination is not quite possible from the observations which can be performed at present.

An alternative approach is to return to equation (52) and try to estimate ρ and measure $H = \dot{R}^*/R^*$, which would then give the sign of k.

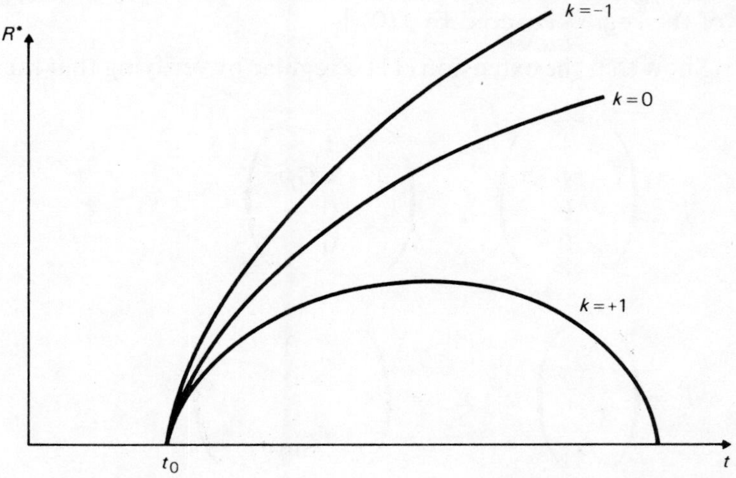

Figure 18. $k=1$: R^* is bounded above so for negative $\ddot{R}^*$ (not→0) the graph must fall to zero. From $\dot{R}^{*2} > -k+d'R^{*-4}$ we have, for $k=-1$: $\dot{R}^* > 1$, so $R^* \to \infty$ faster than t, and for $k=0$: $\dot{R}^* > eR^{*-2}$, so $R^* \to \infty$ faster than $t^{1/3}$.

Unfortunately estimating ρ is very difficult: a lower limit can be placed on ρ by calculating the density of all the types of matter observed—gas, stars, dust, radiation etc.—but it is hard to guess how much non-luminous gas, for instance, passes undetected. Thus it remains an open question whether the universe is finite or infinite in space.

Exercises

1 Find the form taken by the Schwarzschild metric in coordinates u, v, θ, ϕ where u is given by (10) and v is the corresponding coordinate for outgoing null geodesics, i.e.

$$v := -t + r + 2MG \ln|r - 2MG|$$

Now transform to coordinates ξ, η, θ, ϕ where

$$\xi := e^{u/4MG}, \qquad \eta := \begin{cases} +e^{v/4MG} & (r > 2MG) \\ -e^{v/4MG} & (r < 2MG) \end{cases}$$

Evaluate $\xi\eta$ and ξ/η as functions of r and t, and sketch the curves $r = \text{constant}$, $t = \text{constant}$ in the (ξ, η)-plane.

Identify in this plane (a) the region where the metric is regular, (b) the region $r > 2MG$ of the metric (9), and (c) the region covered by the metric (11).

[This is the largest analytic space–time containing the Schwarzschild metric, and it is known as the *Kruskal* metric. In practice, as we have seen, the region outside the matter of a star would only correspond to part of the region covered by (10).]

2 (i) Show that the extension (11) is regular by verifying that the frame

$$\underset{0}{\bar{E}} = \begin{pmatrix} 1 \\ -MG/r \\ 0 \\ 0 \end{pmatrix} \quad \underset{1}{\bar{E}} = \begin{pmatrix} 1 \\ 1 - MG/r \\ 0 \\ 0 \end{pmatrix}$$

$$\underset{2}{\bar{E}} = \begin{pmatrix} 0 \\ 0 \\ r^{-1} \\ 0 \end{pmatrix} \quad \underset{3}{\bar{E}} = \begin{pmatrix} 0 \\ 0 \\ 0 \\ r^{-1}(\sin\theta)^{-1} \end{pmatrix}$$

(in coordinates u, r, θ, ϕ) can be chosen as a freely falling frame at any point.

(ii) Show that, if a vector A has components A^i in (u, r, θ, ϕ)-coordinates, then its components in the above frame are

$$\bar{A}^0 = (1 - MG/r)A^0 - A^1, \quad \bar{A}^1 = A^1 + MGA^0/r$$
$$\bar{A}^2 = rA^2 \quad , \quad \bar{A}^3 = r\sin\theta\, A^3$$

3 A ray of light with tangent vector $X\left(\mathbf{X} = \left(\frac{dt}{ds}, \frac{dr}{ds}, 0, \frac{d\phi}{ds}\right)\right)$ starts to travel outwards $(dr/dt > 0)$ at a point with radial coordinate r, $2MG < r < 3MG$, in the equatorial plane of the Schwarzschild solution. The angle α between the ray and the outward radial direction in the frame $\underset{i}{\bar{E}}$ is defined by $\tan\alpha = \bar{X}^3/\bar{X}^1$. By considering the condition $E^2 > E^2_{\max}$ show that the light will escape if and only if

$$\cot\alpha > \left[\left(\frac{r}{r_0}\right)^2 - \Delta\right]^{\frac{1}{2}} + \frac{1}{3\sqrt{3}}\Delta^{-1}$$

where $\Delta = (1 - 2MG/r)$ and $r_0 = 3\sqrt{3}\,MG$.

4 Show that the spatial metric $dr^{*2} + s(r^*)(d\theta^2 + \sin^2\theta\, d\phi^2)$ takes the form

$$\frac{dr'^2}{1 - kr'^2} + r'^2(d\theta^2 + \sin^2\theta\, d\phi^2)$$

in coordinates r', θ, ϕ where $r'^2 = s(r^*)$ ($= \sin^2 r^*$, r^{*2} or $\sinh^2 r^*$ according as $k = 1, 0$ or -1).

5 An observer starts from rest at radius $r_0 > 2MG$ in the extension (11) of the Schwarzschild metric and falls radially inwards. Show from the geodesic equations that he reaches the singularity at $r = 0$ in a time (i.e. proper time) $s = \pi MG(r_0/2MG)^{3/2}$.
[Note that in conventional units where the velocity of light c is not unity MG must be replaced by MG/c^3 which is equal to about 5×10^{-6}s when M is the mass of the Sun. Note also that the coordinate t (valid only in $r > 2MG$) goes to infinity at $r = 2MG$ (cf. (10)) and so the observer whose journey takes a few microseconds of his own time takes an infinite period of the 'time' of an external observer even to get half way.]

6 (i) Show that a pulse of light emitted from a source at time t_1 and at the spatial origin $(r = 0)$ in the metric (39) will have spread out to a sphere of radius r at time t_2, where

$$r = \int_{t_1}^{t_2} \frac{1}{R(t)}\, dt$$

(ii) Deduce that light emitted from the source during a small time interval

from t_1 to $t_1 + \delta_1$ will be received at radius r over a time interval from t_2 to $t_2 + \delta_2$, where $\delta_2 = \delta_1 R(t_2)/R(t_1)$.
(iii) If the source is emitting N photons per unit time, each with an energy hn, where n is the frequency of the photon at any subsequent stage in its history (see Exercise 2 of chapter 2), show that the brightness of the source (= energy crossing unit area of sphere in unit time) at the radius r is proportional to $R(t_1)^2/(s(r^*)R(t_2)^2 r^2)$. Hence show that, if $R(t)$ is proportional to t^α, then the brightness of a source observed at time t is given, in terms of the coordinate time Δt that has elapsed since its light was emitted, by

$$A\left(\frac{\Delta t}{t}\right)^{-2}\left(1-\alpha\left(\frac{\Delta t}{t}\right)+\mathrm{O}\left(\left(\frac{\Delta t}{t}\right)^2\right)\right)$$

where A is a constant depending on t, the intrinsic brightness (rate of emission of energy) and the parameters of the cosmological model (density etc.).

7 (i) The energy–momentum tensor of a homogeneous and isotropic distribution of electromagnetic radiation satisfies $p = \frac{1}{3}\rho$. Show from (55) that, if this tensor satisfies $T_i{}^j{}_{;j} = 0$ in a homogeneous and isotropic cosmology, then $\rho(t) = \rho_2 (R_2/R(t))^4$ where ρ_2 and R_2 are respectively the pressure and scale-factor at a time t_2.
(ii) If we now suppose that radiation is the dominant constituent of the universe (which was the case at very early times) so that this energy–momentum tensor can be regarded as the total one, show that the equation for $\ddot{R}^*$ can be integrated to give

$$R(t) = R_2\sqrt{(32\pi G\rho_2)}.(t-t_0)^{\frac{1}{2}}$$

where $R(t_0) = 0$.

Index